Nuffield Advanced Mathematics

History of mathematics

LONGMAN

Acknowledgements

The production of this series of advanced mathematics books was made possible by the following grants and support.

1   Generous grants from The Nuffield Foundation and from Sir John Cass's Foundation.

2   Valuable support from
- Northumberland County Council, for seconding Bob Summers to work with the project, and the unstinting help and encouragement of Chris Boothroyd, Director of their Supported Self Study Unit
- David Johnson, Licentiate Professor, Centre for Educational Studies, King's College, London
- the Local Education Authorities of Camden, Essex, Islington, Northumberland, Oxford, Tower Hamlets and Westminster for releasing teachers to work with the project
- Afzal Ahmed, Warwick Evans, Adrian Oldknow and Honor Williams from the Mathematics Centre, West Sussex Institute of Higher Education
- Geoff Wake from the Mechanics in Action Project, University of Manchester.

This course has been written with the help of many people in addition to the team, which consists of: Hugh Neill, Director; Sue Burns, Deputy Director; Sue Ahrens and Bob Summers, Development Officers; and Nina Towndrow, Project Administrator.

Other contributing authors are: Mary Barnes, Jill Bruce, Ruth Farwell, Paul Garcia, Anko Haven, Gerry van den Heuvel-Verhaegh, Alice Hicks, George Gheverghese Joseph, Jan van Maanen, Adrian Oldknow, Tony Ralston, Mary Rouncefield, Paul Tempelaar, Geoff Wake and Peter Wilder.

Teachers who have also helped in the production of materials include: Paul Aljabar, Jane Annets, Dinny Barker, Richard Choat, Joel and Trudy Cunningham, John Deakin, Phil Donovan, Jenny Douglas, John Eyles, Peter Horril, Paul Jenkins, Sue Maunder and Mike Warne.

Other people who have helped: John Fauvel, Steve Russ, Marianne Vinker and Tom Whiteside.

Thanks are given to Mr C B F Walker of The British Museum for his considerable help, and to the University of Oxford Delegacy of Local Examinations for permission to use some examination questions from specimen papers.

The general editor was Hugh Neill. Diona Gregory edited for Longman Education, and kept herself and her team sane and cheerful.

Longman Group Limited
Longman House, Burnt Mill, Harlow, Essex CM20 2JE,
England and Associated Companies throughout the World.

© Nuffield Foundation 1994

First published 1994
ISBN 0582 25729 8

Set in 10/12 Times by Stephen I Pargeter
Illustrated by Hugh Neill
            David Sharp Figure 1.1
Produced by Longman Singapore Publishers Pte Ltd
Printed in Singapore

The publisher's policy is to use paper manufactured from sustainable forests.

Contents

Using Nuffield Advanced Mathematics

- The book consists mainly of activities through which you can develop your understanding of mathematical ideas and results, or apply your knowledge and understanding to problems of various kinds.

- The activities are designed so that you can use your graphics calculator fully, gaining all the advantages that this can bring to your learning of mathematics.

- For most of the activities you can work either individually or in a small group of two to four students, but there are some activities for which individual or group work is specifically recommended.

- You will be working on the activities outside the classroom as well as in class, sometimes with and sometimes without help from your teacher.

- Do not assume that you should work through every activity in a unit. The Nuffield course is suitable for a variety of students with a variety of mathematical backgrounds and interests.

- **Planning your work** on each unit is essential, as you will need to decide together with your teacher, which activities to work on, and whether to work on them in class or as part of your independent study time.

- The information in unit and chapter introductions will help with your planning. The **chapter summaries** and **practice exercises** at the end of the book will also be useful for this purpose. For example, you might already know all of the results that are developed in some particular chapter, in which case it might be appropriate to skip that chapter.

- On several occasions during your study of a unit, you will take part in a **review**. Reviews can take a number of forms: their main purpose is for you to take stock, through discussion with your teacher and other students, of what you have learned so far.

- To assist with the review process, the end of every chapter contains some specific suggestions which might form the basis for discussion; there is also a list headed **What you should know**. This list is a reminder of the main ideas of the chapter and of any new mathematics terms that it has introduced. New mathematical terms are written in **bold** type the first time they are explained in the unit. Most chapters end with some check questions for you to tackle to make sure you have grasped the main ideas of the chapter. There are no answers provided for these questions, but answers are provided for nearly all the other questions on which you will work.

- You will find messages at various stages in the materials suggesting for example that you consult your teacher, or should retain the results of a particular activity for later use. **These will be printed in a different typeface so that you can recognise them.** In addition, there will be general mathematical comments which appear in shaded boxes. These are usually to give you extra background information.

- Towards the back of the book there is a chapter of hints for various activities if you need help in getting started. You will find the hint symbol ◆H◆ in the margin next to the particular question for which there is a hint. You should only look up the hint if you find you are unable to solve the question without it.

Introduction

This book, about the history of mathematics, presents accounts of the development of a number of mathematical topics. These topics have been chosen to relate particularly to your learning of A-level mathematics and to deepen your understanding of such topics as algebra, proof, geometry, calculus and the nature of mathematics. These particular topics form only a small part of what is involved in the study of the history of mathematics. It follows that this book does not attempt to present a coherent story of the development of mathematics.

As you work through activities which illustrate the struggle to express algorithms for solving equations, or to make sense of negative numbers, you may develop a sympathy for present-day students and their difficulties, including your own! You may recognise the puzzlement over 'a minus times a minus makes a plus'. You may also be interested to discover the origins of such terms as algebra, sine or hyperbola.

At the end of this introduction you will find a reference list containing history of mathematics books; we recommend you have at least one available as additional reading. The book by Katz is the most recent and most in sympathy with the aims of this course. In addition, there are collections of source material, available such as the book by Fauvel and Gray.

This book has six units which are designed to be studied in sequence.

First you will study some of the algorithmic materials available in ancient times. The authors have chosen to use Babylonian material; Chinese or Indian material would have been equally possible. Fauvel and Gray state that 'the most startling revaluation this century in the history of mathematics has been the realisation of the scope and sophistication of mathematical activity in ancient Mesopotamia, thanks to the work of scholars in deciphering the cuneiform texts'.

In the second unit, you are introduced to the mathematics of the Greeks who had available to them the mathematics of the ancient orient via trade route connections. For the Greek philosophers, it was no longer sufficient that an algorithm worked; they needed to ask how it worked, and, if possible, to produce rigorous proof that the method worked. In this book the authors concentrate on geometry, in which the Greeks wrestled with, amongst other things, three problems: doubling the cube, trisecting the angle, and squaring the circle. The attempts to solve these problems have resulted in significant mathematical development over the centuries.

During the Greek era, mathematics with a more algorithmic emphasis was continuing to develop, for example, in China and India. In the third unit you will see something of the role of the Arabic-speaking mathematicians in bringing together the mathematics of East and West. In many history books you may find the subsequent contribution of the Arab mathematicians has been understated, and indeed it will be difficult to give a comprehensive evaluation while many Arabic texts remain untranslated. Perhaps one of the major new contributions made by the Arab mathematicians was to use the ideas of Greek geometry to help with the solution of algebraic problems.

Due to its eastern origins, the algebra of the middle ages tended to be algorithmic, compared with the axiomatic foundation of geometry laid by the Greeks. The European algebraists of the 16th century had available to them the Arab and Greek texts, translated via Spanish into Latin. The next great step can be thought of as making the whole field of classical geometry available to the algebraists. In the fourth unit, you will learn how, like the Arab mathematicians, Descartes was interested in the use of algebra to solve geometric construction problems and how he took the extra step of using coordinates to study this relationship.

From Greek times and before, mathematicians had been finding ways to calculate areas bounded by curves and constructing tangents. Solutions had involved clever constructions or algorithms applied to particular curves, but there was no general method. After the development of the analytical geometry of Descartes and his contemporaries, it took only some 40 years more work by a variety of mathematicians, including Newton and Leibniz, to develop what we now know as calculus, a tool by which problems of finding area and tangents can be unified and solved in a general way. In the fifth unit you will see a snapshot of how this came about.

Finally, in the sixth unit, you will be able to follow some of the developments in making mathematics rigorous in the context of the struggle of mathematicians to make sense of negative and complex numbers.

This book is designed to take about 80 hours of your learning time. About half of this time will be outside the classroom.

The units include many mathematicians' names. You need to learn only a few which are detailed in the 'What you should know' lists at the end of each chapter.

Reference list

The crest of the peacock: non-European roots of mathematics, G G Joseph, Penguin, ISBN 0 141 12529 9

A concise history of mathematics, D J Struik, Dover, ISBN 0 486 60255 9

A history of mathematics, V J Katz, Harper Collins, ISBN 0673 38039 4

A history of mathematics (Second edition), by Carl B Boyer and Uta C Merzbac, Wiley, New York 1989, ISBN 0471 50357 6

The history of mathematics: a reader, edited by J Fauvel and J Gray, Open University Press, ISBN 0 33 42791 2

1

The Babylonians

Introduction

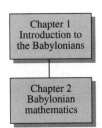

Chapter 1
Introduction to
the Babylonians

Chapter 2
Babylonian
mathematics

In this unit, you will learn how writing and then mathematics developed in the Babylonian era.

Babylon was the major city of one of the world's first civilisations. It was one of the capitals of southern Mesopotamia, formed by the valleys of the Euphrates and Tigris rivers in the area of modern Iraq. The Mesopotamian civilisation flourished for some 3600 years, from 3500 BC until the appearance of the last cuneiform text in AD 75. Mesopotamia was inhabited by the Sumerians from 3500 BC until about 2000 BC, and then by the Babylonians from 2000 BC until AD 75. Many of the features of modern society – complex social organisation, political and religious structures, urban centres, sophisticated productive and economic systems, for example – first developed in the four great river valley civilisations of the time. The other three were around the Nile, the Indus, and the Yellow rivers. Such features depended on development in the two basic human languages: words and numbers. Our writing and mathematics are rooted in Mesopotamia.

The unit begins with some background introduction. Chapter 2, which is far more mathematical, is concerned with the ways in which the Babylonians carried out arithmetic and mathematics using a number system based on 60. They were able to find square roots and to solve quadratic equations.

> This unit is designed to take about 10 hours of your learning time. About half of this time will be outside the classroom.
>
> Read Chapter 1 quickly. Its purpose is to give you a background to Babylonian mathematics. Then go on to Chapter 2.
>
> There are summaries and further practice exercises in Chapter 12.

Mathematical knowledge assumed

- knowledge of quadratic equations would be useful.

1 Introduction to the Babylonians

This chapter gives you important background to the society of the Babylonians, and the context in which they developed their mathematics.

There are no mathematical activities in this chapter.

Read this chapter as background, so that you can understand the achievements described in Chapter 2.

If you would like more background information, a good reference is *Early Mesopotamia* by J N Postgate, Routledge, ISBN 0415 110 327.

The map in Figure 1.1 shows you where the Babylonians lived, around 2000 BC until AD 75.

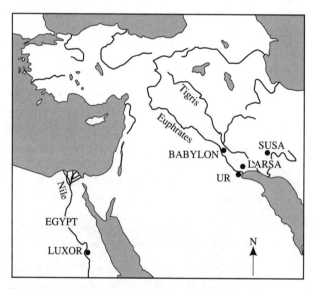

Figure 1.1

Background to the Babylonians

The following extracts give a general impression of Mesopotamian life in the second millennium BC. They show you something of a culture whose needs had previously led to the development of writing.

Legal dispute about a lost donkey

This extract is about a pack-ass which Ilsu-abusu hired from Warad-Enlil and Silli-Istar, his brother, in the town of Simurrum, and then lost.

> On the subject of the pack-ass Ilusu-abusu, Warad-Enlil and Silli-Istar went to law, and the judges dispensed justice to them in the Gate of Samas inside the city of Sippar. On behalf of Ilusu-abusu the judges committed Warad-Enlil and Silli-Istar to (swear by) the emblem of Samas. Without an oath-taking Ilusu-abusu the son of Sin-nasir, Warad-Enlil and Silli-Istar reached agreement at the emblem of Samas in the Gate of Samas. Because he did take their donkey, Ilusu-abusu shall not make any claim against Warad-Enlil and Silli-Istar for the 6 shekels of silver (paid) at Zaban (near the Diyala) of the 10 shekels of silver (paid) in Greater Sippar.

A murder trial at Nippur

> Nanna-sig, Ku-Enlila the barber, and Enlil-ennam slave of Adda-kalla the gardener killed Lu-Inana the priest. After Lu-Inana's death, they told Nin-dada, wife of Lu-Inana, that Lu-Inana her husband had been killed. Nin-dada, opened not her mouth, covered it up. Their case was taken to Isin before the king. King Ur-Ninurta ordered their case to be taken for trial in the Assembly of Nippur.
>
> Ur-Gala, Dudu the bird-catcher, Ali-ellati the muskenum, Puzu, Eluti, Ses-kalla the potter, Lugalkam the gardener, Lugal-azida, and Ses-kalla took the floor and stated: 'They are men who have killed a man; they are not live men. The three males and that woman should be killed before the chair of Lu-Inana son of Lugal-urudu the priest.'
>
> Suqallilum the ..., soldier of Ninurta, and Ubar-Sin the gardener took the floor and stated: 'Did Nin-dada in fact kill her husband? What did the woman do that she should be put to death?'
>
> The Assembly of Nippur took the floor and stated: 'A man's enemy may know that a woman does not value her husband and may kill her husband. She heard that her husband had been killed, so why did she keep silent about him? It is she who killed her husband, her guilt is greater than that of the men who killed him'.
>
> In the Assembly of Nippur, after the resolution of the case, Nanna-sig, Ku-Enlila, Enlil-ennam and Nin-dada were given up to be killed. Verdict of the Assembly of Nippur.

Assur merchant explains lack of textiles to his Kanes partners

> As to the purchase of Akkadian textiles, about which you wrote to me, since you left the Akkadians have not entered the City (of Assur). Their country is in revolt. If they arrive before winter, and there is the possibility of a purchase which allows you profit, we will buy for you and pay the silver from our own resources. You should take care to send the silver.

Letter to Hammurabi's governor at Larsa

> Say to Samas-hazir, Enlil-kurgalani says: May Adad keep you alive! About the field of Ahum-waqar: As you know, he has been enjoying the use of the field for 40 years, and now he is going on one campaign in the king's corps,

> but Sin-imguranni has now taken the field away from him and given it to a servant of his. Look into the matter and don't allow him to suffer an injustice.

About 1750 BC Hammurabi developed Babylon into a powerful city-state controlling a regional empire. Although it had a large army and was well governed, Babylonia relied for survival on precarious alliances and coalitions with neighbouring states. Hammurabi's aims were to acquire more land and to ensure that vital raw materials became available to Babylonia, which, though very fertile, had few other natural resources.

With its strong central government, Babylonia at first prospered from taxes and payments made by dependent states. This wealth financed extensive state irrigation and building projects. Then, shortly after his death in about 1750 BC, distant provinces broke away, and the central government declined.

However, the nation-state of Babylonia gained lasting historical significance. Babylon became the leading city of the land, the capital of Mesopotamia.

The most important written document of this time was Hammurabi's law code.

Here are some fragments of his law code.

From the prologue to the Code of Hammurabi

> When exalted Anum, king of the Anunnaki (the major gods) and Enlil, lord of heaven and earth, who decrees the destinies of the Land, decreed for Marduk, first-born son of Enki, the role of Enlil to the entirety of people, making him the greatest of the Igigi-gods, called Babylon by its exalted name, made it pre-eminent in the world, and established in it a permanent kingship whose basis is founded like heaven and earth …

> … at that time Anum and Enlil called my name to improve the living conditions of the people, Hammurabi, attentive prince, revering the gods, to make justice appear in the land, to abolish the criminal and evil, to stop the strong from oppressing the weak, to rise like Samas (Sun-god, God of Justice) over mankind and to illuminate the land.

Code of Hammurabi, paragraph 5

> If a judge tried a case and made a decision and had a sealed document executed, but later changed his judgement, they shall convict that judge of changing his judgement: he shall pay twelve times the claim involved in that case, and they shall remove him in the Assembly from his judgement seat, and he shall not sit in judgement with judges again.

Code of Hammurabi, paragraph 32

> If a merchant has ransomed a soldier or a fisherman who was taken captive on a campaign of the king, and enabled him to regain his village, if there is enough to ransom him in his household, he shall pay his own ransom; if there is not enough in his household to ransom him, he shall be ransomed from (the resources of) the village temple; and if there is not

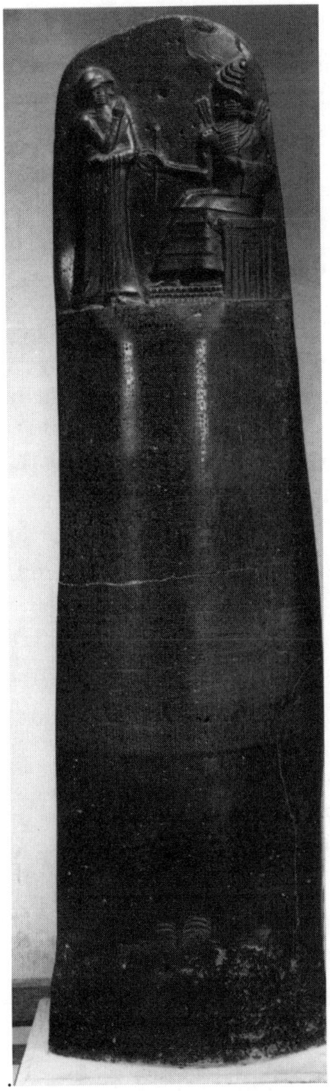

enough in his village temple to ransom him, the palace shall ransom him. His field, orchard or house shall not be sold for his ransom money.

Code of Hammurabi, paragraph 36/7

The field, orchard, or house of a soldier, a fisherman, or a (palace) tenant shall not be sold.

If a man buys the field, orchard or house of a soldier, fisherman, or (palace) tenant, his tablet shall be broken and he shall forfeit his silver. The field, orchard or house shall return to its owner.

Code of Hammurabi, paragraph 55/6

If a man has opened his channel for irrigation, and has been negligent and allowed the water to wash away a neighbour's field, he shall pay grain equivalent to (the crops of) his neighbours.

If a man has released the water and allowed the water to wash away the works (i.e. furrows and soil preparation) on a neighbour's field, he shall pay 10 gur of grain per bur.

Code of Hammurabi, paragraph 60/1

If a man has given a field to a gardener to plant an orchard, and the gardener planted the orchard, he shall grow the orchard for four years, and in the fifth year the owner of the orchard and the gardener shall divide equally, the owner of the orchard choosing his share.

If the gardener did not complete planting the field but left bare ground, they shall put the bare ground into his share.

Code of Hammurabi, paragraph 144/5

If a man married a naditum and that naditum has given a slave-girl to her husband and (so) produced sons, and that man decides to marry a concubine, they shall not consent to that man, he shall not marry the concubine.

If a man married a naditum and she has not got sons for him, and he decided to marry a concubine, that man may marry the concubine: he may bring her into his house, but she shall not be made the equal of the naditum.

Code of Hammurabi, paragraph 196/9

If a man has put out the eye of another man, they shall put out his eye.

If he breaks the bone of another man, they shall break his bone.

If he puts out the eye of a muskenum, or breaks the bone of a muskenum, he shall pay 1 mina of silver. If he puts out the eye of the slave or another man, or breaks the bone of the slave or another man, he shall pay half his price.

Figure 1.2
Stele inscribed with Hammurabi's law code, excavated at Susa in 1900. The text is inscribed in the traditional vertical columns in the cuneiform script. Above, Samas, seated on his altar and with his solar rays springing from his shoulders, commissions Hammurabi to mete out justice to the people.

Evolution of counting and writing

A group of Old Sumerian clay tablets may be some of the world's earliest forms of written communication, for they probably pre-date Egyptian hieroglyphic writing. The tablets, each about the size of the palm of a hand, were found in Uruk, southern Mesopotamia, and date from about 3200 BC. The Sumerians used a round stylus to make the circular and D-shaped impressions that you can make out on these tablets, and a pointed stylus to incise the pictures. A tablet and styluses are shown in Figure 1.3.

What might these signs have represented? What were the tablets used for? How did writing come into being? The answers to these questions can be traced back to earlier developments of counting and using symbols.

Throughout most of human history people have lived without writing; yet images, symbols and other devices were used to transmit thought long before people could write. During the Upper Palaeolithic era (the Old Stone Age), about 30 000 to 12 000 years ago, incised bones may have been used to keep track of lunar time. If so, these pieces of bone represent the earliest known system of counting, or 'reckoning'. Most such developments, however, ended with the decline of Palaeolithic cultures by the end of the last ice age, about 10 000 BC.

Throughout the Near East, small clay or (less often) stone tokens of varying shapes, which date back to the eighth millennium BC, have surfaced in archaeological excavations.

Figure 1.4 shows these small – 3 mm to 2 cm – symbolic tokens which may represent counters in an archaic recording system. It is believed that some of these tokens represented a measure of some commodity.

Figure 1.3a

Figure 1.3b

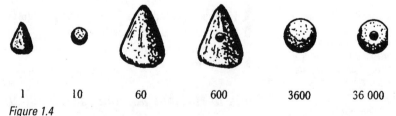

| 1 | 10 | 60 | 600 | 3600 | 36 000 |

Figure 1.4

Figure 1.5 shows other tokens which might have represented commodities.

The tokens would have been used for stock-taking, or for keeping records of transactions, or for bartering, or conceivably for tabulating the number of animals in a flock or the produce from the harvest.

This token system remained in place from the eighth to the third millennium BC.

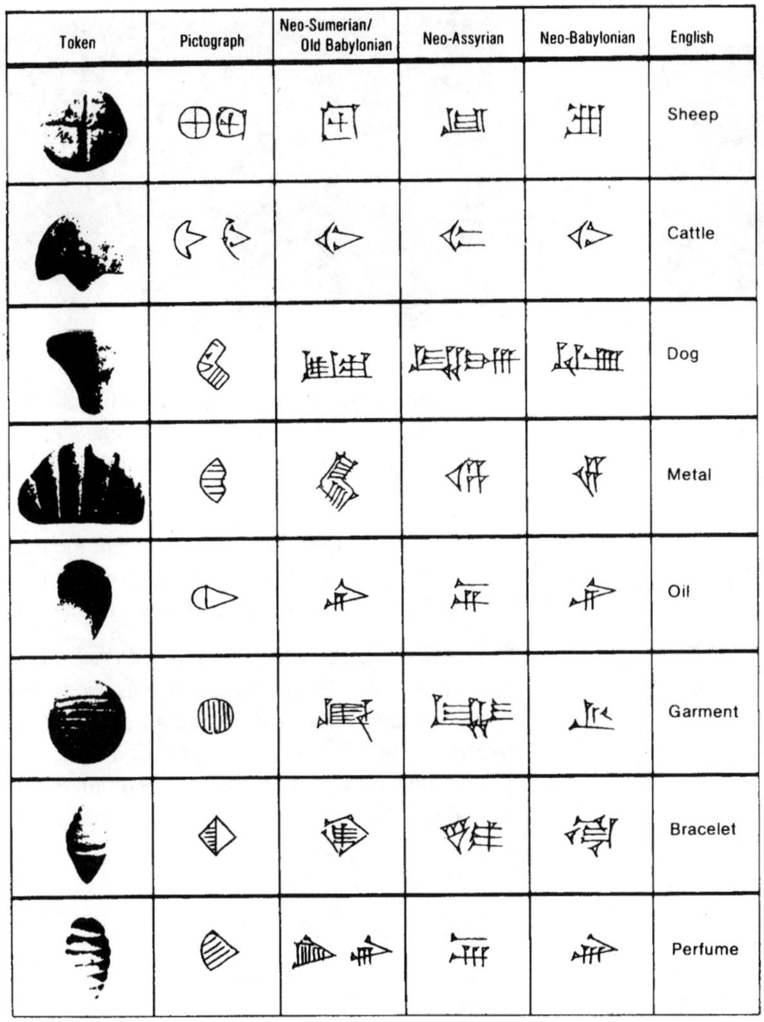

Token	Pictograph	Neo-Sumerian/ Old Babylonian	Neo-Assyrian	Neo-Babylonian	English
					Sheep
					Cattle
					Dog
					Metal
					Oil
					Garment
					Bracelet
					Perfume

Figure 1.5

The next stage in the evolution of writing occurred shortly after 3500 BC, during the Uruk period. Commercial exchange within and among the earliest cities required intermediaries, who may have used a small spherical clay envelope – called a bulla – to keep or to transport the tokens, which represented specific transactions such as buying or selling ten loaves of bread or sixty measures of oil.

The bulla, which was approximately the size of a tennis ball, was sealed to prevent tampering with the tokens. It could also be impressed with a cylindrical seal, such as that shown in Figure 1.6, to show ownership.

Figure 1.6a

Figure 1.6b

A drawback of these opaque clay envelopes was that they concealed the tokens from view, so they could not be counted or checked without destroying the bulla. So people began to impress the surface of the fresh, pliable clay ball with the number and rough shape of the tokens within.

These marks are the crucial link between the three-dimensional token system and a two-dimensional writing system. Anyone familiar with the tokens could now 'read' the clay balls; the tokens themselves soon became superfluous.

The earliest tablets, which supplanted the bullae but resembled them in size, shape and even convexity, also functioned as economic records. In time, the signs on the tablets became simplified, and small drawings of the tokens were incised with a pointed stylus. Although the signs used on the earliest tablets were not actual pictographs, newer words came to be depicted by true pictographs in the likeness of the objects that they represented, as shown in Figure 1.7.

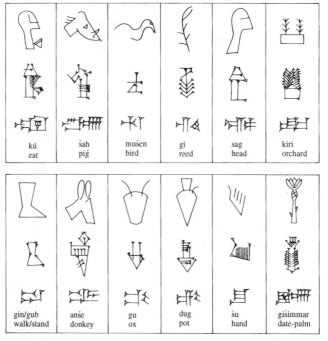

Figure 1.7

A different development in early Mesopotamian writing was the 'rebus principle'. Consider, for example, the English word 'treaty', which is difficult to show in a picture or pictograph. The rebus representation of treaty might be a 'tree' plus a 'tea' cup or a 'tea' bag. The rebus principle simplified writing, reduced ambiguity, and increased the range of expressible ideas. The ultimate consequence of this rebus writing was the development of phonetics – the representation of the sounds that express an idea, rather than depiction of the idea itself.

The Mesopotamians accomplished this evolution by about 2500 BC.

They had reduced the number of signs from over 2000 to about 600. By adopting a special stylus, usually made from reed and shown in Figure 1.8, they had given the Sumerian script its standardised cuneiform, wedge-shaped, character.

Figure 1.8

The earliest tablets were purely economic documents, but cuneiform writing came eventually to express political, judicial, literary, philosophical, religious, and scientific ideas. The cuneiform system of writing continued in use for almost 3000 years until Aramaic, an alphabetic writing system, replaced it.

Reflecting on Chapter 1

What you should know

- the general circumstances in which the Sumerians and Babylonians invented their mathematics
- how to explain on what evidence our knowledge of the Babylonians is based.

Preparing for your next review

- Be ready to explain orally the context behind Babylonian mathematics.
- Write a brief account of the place of mathematics in Babylonian society in about 1800 BC.
- Write brief notes on the importance of trade for the Sumerian and Babylonian development of a written number system.

There are no practice exercises for this chapter.

2 Babylonian mathematics

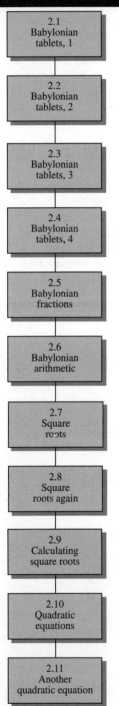

In this chapter you will learn how the Babylonians represented and wrote their numbers, and how they carried out calculations. You will also learn that they used the result now known as Pythagoras's theorem, and could calculate square roots and solve quadratic equations.

Activities 2.1 to 2.5 introduce you to the Babylonian number system and the tablets used as evidence of Babylonian mathematics.

Activity 2.6 is an introduction to Babylonian arithmetic. You can find more detail in, for example, *The crest of the peacock*, by G G Joseph.

Activities 2.7 to 2.9 are about the way Babylonians found square roots. This develops in Activities 2.10 and 2.11 to the solution of quadratic equations.

In Activities 2.12 and 2.13, you will learn something about the geometry which the Babylonians knew.

The activities are designed to be worked in sequence.

Activities 2.12 and 2.13 are optional.

All the activities are suitable for working in a group.

About 500 of the half million or so inscribed clay tablets excavated are thought to be of mathematical interest. These tablets are now scattered around the museums of Europe, Iraq, and American universities. The Babylonian mathematical texts were largely undeciphered until the 1930s. Even now evidence continues to be uncovered.

A sexagesimal number system is a system based on 60 as opposed to the decimal system which is based on 10.

Figure 2.1

The shapes in Figure 2.2 are stylised versions of the tokens shown in Figure 1.4.

The mathematical tablets appear to originate from three main periods. As you have seen, the oldest tablets go back to the Old Sumerians and date from about 3200 BC. The mathematics on these tablets tends to consist of commercial information, and shows a developing sexagesimal system.

The majority of the tablets appear to belong to what is known as the Old Babylonian period, around the time of the reign of Hammurabi.

The newest tablets have been dated at around 300 BC, from the New Babylonian period which followed the destruction of Nineveh. These tablets contained a lot of astronomical data.

In this chapter, after a brief look at an Old Sumerian tablet, you will concentrate on the work of the Old Babylonians.

The number system on the Babylonian clay tablets is based on the Old Sumerian system, but there are some differences.

The Old Sumerian clay tablet shown in Figure 2.1 dates from about 3200 BC and records a transaction. The Sumerians used symbols such as those shown in Figure 2.2.

1	10	60	600	3600
D	o	D	◎	○

Figure 2.2

The symbols for 1 and 10 are made with a small stylus, held obliquely for 1 and vertically for 10. The symbols for 60 and 3600 are made similarly, but with a larger stylus. However, the Sumerians did not use any form of place value.

Activity 2.1   Babylonian tablets, 1

Figures 2.3a and 2.3b show transcriptions of the front (obverse) and back (reverse) of an Old Babylonian clay tablet containing mathematical material.

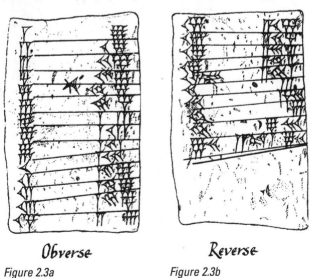

Obverse

Figure 2.3a

Reverse

Figure 2.3b

1 **a** Decipher the table and convert it to decimal notation, disregarding the last line of the back.

b What kind of table is it?

2 Describe the structure of the Babylonian number system.

3 Look back to Figure 2.1. The top third of the tablet represents one number. What number is it?

4 What is the main difference between the Old Sumerian and the Babylonian number systems?

Activity 2.2   Babylonian tablets, 2

Figure 2.4a shows a transcription of a similar tablet found in Susa.

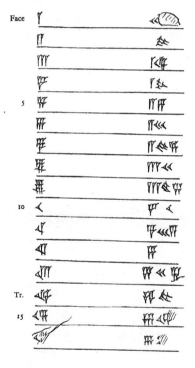

Figure 2.4a   *Figure 2.4b*

1 Decipher the table shown in Figure 2.4a. What kind of table is it?

2 The tablet is damaged, making the numbers in the first line and the last two lines hard to recognise. Re-construct these lines.

3 Compare the fifth number in the left-hand column with the twelfth number in the right-hand column. Write a brief explanation of your findings.

Figure 2.4b shows a tablet whose size is about 8 cm by 5 cm (and about 2 cm thick), on which the multiplication table for ten is written.

Figure 2.5

4 Transcribe and re-construct the middle and right-hand columns of Table 2.4b (the left-hand column consists of multiplication signs).

During the Old Babylonian age, the inhabitants of Mesopotamia achieved a high level of civilisation which carried out complex operations like paying salaries, determining the yield of land, dividing a legacy, calculating compound interest, and so on. Such economic actions required considerable reckoning skill. Many tablets showing various kinds of mathematical tables indicate that the Babylonians possessed such skills.

The first nine lines from one of these tablets are transcribed in Figure 2.5.

Activity 2.3　　*Babylonian tablets, 3*

1 Decipher the table shown in Figure 2.5.

 2 How do the numbers in each line relate to one another?

 3 Re-interpret the sixth and seventh numbers of the right-hand column.

Although you may not be entirely clear about the meaning of the table, you can probably conclude that a sign, which indicates the start of the fractional part of a number, is missing.

Suppose that the Babylonians were not familiar with a so-called sexagesimal point.

4 What does the table probably mean? And for what can it be used?

5 Two numbers have been omitted in the left-hand column. Why?

Babylonian numbers

The Babylonian number system is sexagesimal and positional: it uses only two numerals, $\mathbf{7}$ and $\blacktriangleleft$, which stand for one and ten. The sexagesimal feature means that the numbers from 1 to 59 are composed of repetitions of the two numerals. Thus, for example, the number 47 consists of 4 tens and 7 units.

The positional, or place value, feature means that $\mathbf{7}$ can also represent 1×60^n and $\blacktriangleleft$ can represent 10×60^n where n can take any integral value – either positive or zero for whole numbers, or negative for fractions. The correct value of n in any particular case must be deduced, or guessed, from the context.

Another ambiguity in the Babylonian number system occurs because you cannot always recognise the number of sexagesimal positions within a number. For example, the number $\mathbf{77}$ can consist of either one or two sexagesimal positions.

Activity 2.4　　*Babylonian tablets, 4*

1 What might the numbers $\mathbf{77}$ and $\blacktriangleleft\mathbf{7}\blacktriangleleft$ mean?

Until almost 300 BC the Babylonians had no clear way to indicate an 'empty' position; that is, they did not have a zero symbol, although they sometimes left a space where a zero was intended.

2 Write the numbers 156 and 7236 in Babylonian notation.

3 What might the number ⵖⵖ mean now?

You can conclude that the Babylonian invention of the zero – possibly the oldest zero in history, which they wrote as an empty space   or ⵖ – is a logical result of their positional number system.

The zero was introduced to indicate an empty position inside a number, or at the beginning of a number, but it was not used at the end of a number.

In spite of its deficiencies, the sexagesimal system used by the Babylonians offered many advantages. The number 60 has many factors which make it good for dealing with fractions, and its large size makes it useful for dealing with large numbers. It was used by Greek and Arab astronomers for many centuries. This is why a sexagesimal system is still used to measure time and angles.

4 Decipher the Larsa tablet shown in Figure 2.6. Suggest what the cuneiform words (columns two and four) might mean.

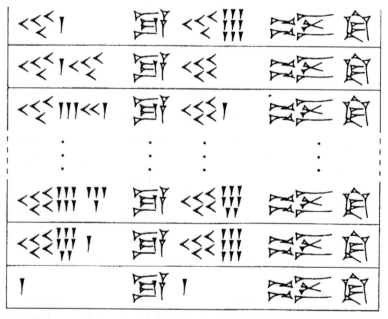

Figure 2.6

Number notation

The Babylonian number system has some ambiguities which arise from the lack of a sexagesimal point and of a symbol for zero, as well as from the fact that you cannot always recognise sexagesimal positions. It is convenient to transcribe Babylonian numbers into a more readable form, by writing a semicolon to separate the integral and fractional part of a number, a comma to distinguish the several sexagesimal positions, and a zero where it is intended.

Activity 2.5   Babylonian fractions

1 Write the sexagesimal numbers 2,0,48;12 and 3;9,18 in decimal notation.

2 Convert the decimal numbers 10 000 and $60\frac{3}{5}$ into sexagesimal numbers.

Not all fractions are as easy to convert.

 3 Convert $\frac{1}{9}$ and $\frac{1}{27}$ into sexagesimal numbers.

In Activity 2.3, question **5**, the 'irregular' numbers 7 and 11 were omitted from the table of reciprocals because they are non-terminating in a sexagesimal system, similar to $\frac{1}{3}$ in our decimal system.

4 In the decimal system, which fractions convert to terminating decimals and why? Examine the connection between the denominators of the fractions and the number ten.

5 a Which fractions, when converted to sexagesimals, have representations which terminate?
b What may be the advantage of the sexagesimal system when dealing with fractions?

6 Write in sexagesimal notation those of the following fractions which have representations which terminate: $2\frac{3}{22}$, $6\frac{5}{32}$ and $83\frac{1}{72}$.

By considering non-terminating fractions, the Babylonians were faced by the problem of infinity. However, they did not examine it systematically. Not until the Greeks would this challenge be tackled.

At one point, a Mesopotamian scribe seemed to give upper and lower bounds for the reciprocal of the irregular number 7, placing it between 0; 8,34,16,59 and 0; 8,34,18.

7 Verify the bounds above by giving the periodicity of $\frac{1}{7}$ in sexagesimal notation when dealing with fractions.

Except for the lack of a separator between the integral and fractional part of a number, and a symbol for zero, the Babylonian number system seems to be equivalent to the modern decimal system. It enabled the Babylonians to do arithmetic operations much as you calculate today, and with comparable facility.

Some of these operations can be found on an Old Babylonian tablet whose transcription is depicted in Figure 2.7. It contains text, numbers and operators, and describes a method for solving a problem which is composed of two simultaneous equations – that is, a 'length' and a 'width' which you would express as x and y.

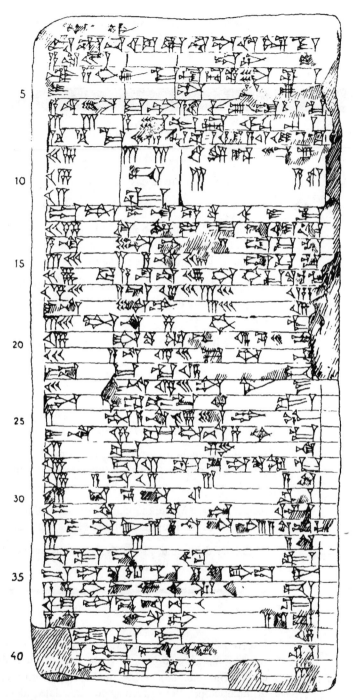

Figure 2.7

Later in this chapter, you will examine more closely how the Babylonians dealt with such problems and how they solved quadratic equations.

But first, in Activity 2.6, you will concentrate on their arithmetic operations.

1 Referring to Figure 2.7, find and check the answers of the following sums and reproduce also the 'signs' for addition and multiplication.

a $14,30 \times 14,30$ (line 17)

b $0,30 + 14,30$ (lines 21, 22)

c 15×12 (line 29)

Summation, subtraction, and even multiplication seem to have been carried out in the same way as we handle them today.

However, to be able to multiply, the Babylonians had to have multiplication tables containing the products 1×2, 2×2, ... , 59×2; 1×3, 2×3, ... , 59×3; 1×59, 2×59, ... , 59×59. In practice, the multiplication table for nine, for example, only consisted of 1×9, 2×9, ... , 19×9, 20×9, 30×9, 40×9, 50×9, as you saw in Figure 2.4a. Division was carried out by multiplying by the reciprocal of the divisor.

Thus, to calculate $47 \div 3$, you first determined $1 \div 3$ (by looking it up in a table for reciprocals) and then multiplied the result by 47.

Besides a table of reciprocals, which contained the reciprocals from 1 to 81 of those fractions giving terminating sexagesimals, the Babylonians also used tables containing multiples of these reciprocals for multiplication.

For example, here is the multiplication table for 0; 6,40 ($1 \div 9$).

$1 \times 0;6, 40 = 0;6, 40$

$2 \times 0;6, 40 = 0;13, 20$ $\qquad$ $19 \times 0;6, 40 = 2;6, 40$

$3 \times 0;6, 40 = 0;20$ $\qquad$ $20 \times 0;6, 40 = 2;13, 20$

$4 \times 0;6, 40 = 0;26, 40$ $\qquad$ $30 \times 0;6, 40 = 3;20$

$5 \times 0;6, 40 = 0;33, 20$ $\qquad$ $40 \times 0;6, 40 = 4;26, 40$

$6 \times 0;6, 40 = 0;40$ $\qquad$ $50 \times 0;6, 40 = 5;33, 20$

$7 \times 0;6, 40 = 0;46, 40$

$8 \times 0;6, 40 = 0;53, 20$

These tables were used to carry out division.

2 Find an example of a subtraction in the top part of Figure 2.7.

3 **a** Explain how to divide 47 by 9, using the Babylonian method.

b Divide 17, 9 by 64, using the Babylonian method.

Measuring the diagonal of a square

Figure 2.8 shows an Old Babylonian tablet, whose diameter is about 7 cm, with its transcription. The tablet shows a square with its diagonals, and seems to contain three numbers. It is actually the equivalent of a student's exercise book.

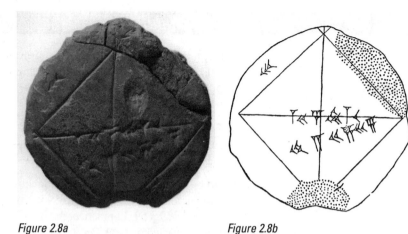

Figure 2.8a   Figure 2.8b

Activity 2.7   Square roots

1 Decipher the numbers on the tablet (Figure 2.8b).

2 What do these numbers mean?

3 How is the number under the diagonal related to the one above?

4 At one point the number under the diagonal is damaged. What symbol is missing?

But what was the student doing? The number on the diagonal is the length of the diagonal of a square of side 1 unit. The given square is of side $\frac{1}{2}$ unit, so the number underneath the diagonal is the result of multiplying the diagonal by $\frac{1}{2}$ to obtain the diagonal of a square of side $\frac{1}{2}$. In fact, it is a scaling operation.

Activity 2.8   Square roots again

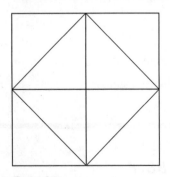

Figure 2.9

You may wonder how the Babylonians knew that the length of the diagonal is $\sqrt{2}$. They could have found this either by geometrical study, or by applying the result which was later to be called Pythagoras's theorem, which they probably knew, though not by that name!

1 Deduce from Figure 2.9, without using Pythagoras's theorem, that the length of the diagonal of the square is $\sqrt{2}$.

It is possible that Pythagoras later arrived at his theorem by considering this kind of diagram.

2 Use Figure 2.9 to prove Pythagoras's theorem in an isosceles right-angled triangle.

3 The diagram in Figure 2.10 is based on a Chinese text from between 1100 and 600 BC. Prove Pythagoras's theorem from this diagram.

Figure 2.10

Pythagoras's theorem might have appeared, and even been proved in a way similar to your method above, although this is not known for certain. What is known is that other tablets from the Old Babylonian age contain calculations which seem to be based on Pythagoras's theorem or to have involved Pythagorean number triples. This would imply that the Babylonians were already familiar with the result more than a thousand years before Pythagoras (circa 500 BC)!

Later in the chapter there is a section about Pythagoras's theorem.

Return to the $\sqrt{2}$ approximation. You can check the accuracy of this approximation by squaring 1; 24,51,10. You will find 1; 59,59,59,38,1,40 which corresponds to an error of less than $\dfrac{22}{60^4}$.

If you express this as a decimal fraction, you obtain the very good approximation of 1.414 213 ... for $\sqrt{2}$. A more accurate approximation is $\sqrt{2} \approx 1.414\ 214$ ... .

4 Devise an algorithm which would give you an approximation to this degree of accuracy.

A square root algorithm

How did the Babylonians achieve such accuracy for the square root of two? It seems likely that they used an iterative method along the following lines.

Activity 2.9 Calculating square roots

$\sqrt{2}$ indicates the side of a square with area 2. Start with a rectangle of area 2 and sides, for example, 1 and 2. Figure 2.11 illustrates the following method to obtain a rectangle of area 2 which is closer to a square.

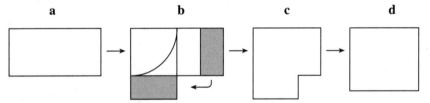

Figure 2.11

Taking a corner of the rectangle as centre, and the smaller side as radius, draw a quadrant of a circle. Mark off the square in the corner of the rectangle. Bisect the remaining rectangle, and slide the shaded piece into the new position shown. The outline of the new shape is shown in Figure 2.11c. Take the longer side of this nearly square shape in Figure 2.11c as your new approximation to the longer side of the rectangle in Figure 2.11d; then find the shorter side so that the area of this rectangle is 2. Now start the procedure again with this new rectangle, which is more nearly a square than the one with which you started.

1 a What are the dimensions of your first new rectangle in this case?

b Find out how many iterations you need to reach the approximation 1; 24,51,10 for $\sqrt{2}$.

2 Experiment with the algorithm to see how you can find other square roots. What difference does it make when you start with other values, such as 1 and 5, or 2 and $2\frac{1}{2}$ for $\sqrt{5}$?

3 How does this algorithm compare to other square-root algorithms that you know?

4 Use this algorithm to find a formula for approximating to $\sqrt{a^2 + h}$.

Quadratic problems involving one unknown

The existing mathematical tablets from the Old Babylonian period fall broadly into two categories: mathematical tables, and problems. You have seen examples of tables in the section about number notation: these include multiplication tables, reciprocal tables, and the table of squares. These tablets consist solely of tables of numbers; about four hundred such tablets have been found.

Problem tablets, by contrast, are rarer – only a hundred or so of them have been found. Below are two examples. Notice that, although the words 'side' and 'square' are used, the problem is not really geometrical, unless Babylonian geometry was quite different from ours.

The Babylonians probably used the terms 'side' and 'area' to mean the unknown quantity and its square, rather than geometric, quantities. This is similar, perhaps, to the way in which you think of x squared without necessarily imagining a square. Yet it is possible that the terms 'side' and 'area' trace back to an earlier geometric use. This is in contrast to the algebra of the Arab mathematicians for whom there was a strong association between computational procedures and geometric representation. This association remained until the time of Descartes in western mathematics.

> I have added up the area and the side of my square: 0; 45. You write down 1, the coefficient. You break off half of 1. 0; 30 and 0; 30 you multiply: 0; 15. You add 0; 15 to 0; 45: 1. This is the square of 1. From 1 you subtract 0; 30, which you multiplied. 0; 30 is the side of the square.

In that example, and in the next, the problem is given in the first sentence; the rest is its solution. Here is the second problem.

> I have subtracted the side of my square from the area: 14,30. You write down 1, the coefficient. You break off half of 1. 0; 30 and 0; 30 you multiply. You add 0; 15 to 14,30. Result 14,30; 15. This is the square of 29; 30. You add 0; 30, which you multiplied, to 29; 30. Result 30, the side of the square.

Activity 2.10   Quadratic equations

1 Verify that the results 0;30 and 30 solve the problems quoted above.

2 Express the first problem in modern algebraic notation.

3 In modern algebraic notation follow the instructions for the first problem. Call the unknown side, x, the coefficient of x, (which here is 1) b, and the number in the statement of the problem (here 0; 45) c. The problem is then to find x, where $x^2 + bx = c$.

4 Follow the instructions for solving the second problem using modern algebraic notation. Call the unknown side x, the coefficient of x (which here is 1) b, and the number in the statement of the problem (here 14,30) c. The problem is then to find x, where $x^2 - bx = c$.

5 Is the algorithm familiar to you?

6 Deduce the solution of the equation $x^2 - bx = c$ by completing the square.

7 How does this algebraic deduction compare with what the Babylonian scribe did? Approach this by considering
a the similarities
b the differences
between the two approaches.

8 Solve the equation $11x^2 + 7x = 6;15$ by first multiplying through by 11, and then making the transformation $y = 11x$.

Question **8** of Activity 2.10 illustrates the use of algebraic transformations in Babylonian mathematics.

Quadratic problems involving two unknowns

The Babylonians had algorithms for solving quadratic equations which fall into three types, shown in Table 2.1.

In Activity 2.10, you saw examples of both the first and the second type.

The Babylonian way of solving the first equation is identical to al-Khwarizmi's algorithm which you will consider in Unit 3 on the Arab mathematicians.

In this section you will study the third type of equation.

Type	Quadratic
1	$x^2 + ax = b$
2	$x^2 - ax = b$
3	$\begin{cases} x+y = a \\ xy = b \end{cases}$

Table 2.1

1 a Before considering the Babylonian algorithm for solving the third type of equation, reflect on the system of equations $\begin{cases} y - x = a \\ xy = b \end{cases}$. Re-write this system as one equation, a quadratic in x.

b Why is it not necessary to include the system $\begin{cases} y - x = a \\ xy = b \end{cases}$ in the list of three types given in Table 2.1?

An example of the system $\begin{cases} y - x = a \\ xy = b \end{cases}$, where $xy = 1,0$ (60 in decimal), is given in Figure 2.12. The Akkadian terms '*igūm*' and '*igibūm*' refer to a pair of numbers which stand in the relation to one another of a number and its reciprocal, to be understood in the most general sense as numbers whose product is a power of 60.

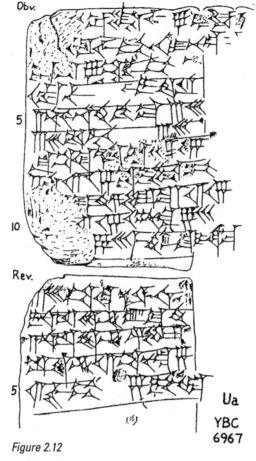

Figure 2.12

Obverse

1	[The *igib*]*ūm* exceeded the *igūm* by 7.
2	What are [the *igūm* and] the *igibūm*?
3-5	As for you - halve 7, by which the *igibūm* exceeded the *igūm*, and (the result is) 3;30.
6-7	Multiply together 3;30 with 3;30, and (the result is) 12;15.
8	To 12;15, which resulted for you,
9	add [1,0, the produ]ct, and (the result is) 1,12;15.
10	What is [the square root of 1],12;15? (Answer:) 8;30.
11	Lay down [8;30 and] 8;30, its equal, and then

Reverse

1-2	subtract 3;30, the *takiltūm*, from the one,
3	add (it) to the other.
4	One is 12, the other 5.
5	12 is the *igibūm*, 5 the *igūm*.

Ua
YBC
6967

2 Follow through the instructions in a modern algebraic format.

3 How should you solve the pair of equations $\begin{cases} x + y = 6;30 \\ xy = 7;30 \end{cases}$ if you were a Babylonian scribe?

Geometry in the Old Babylonian period

So far you have explored the algebra of the Old Babylonians. Until recently their contribution to geometry was thought to be much less than is now known to be the case. In particular, a group of mathematical tablets, which seems to challenge this assumption, has been found at Susa. In Activity 2.12, you will see an example of one such tablet.

 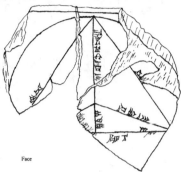

Figure 2.13a   Figure 2.13b

The Old Babylonian tablet shown in Figure 2.13 was found in Susa, and appears to show an isosceles triangle and its circumscribed circle.

The number on the base of the triangle is 60, and the numbers on the equal sides are 50.

1 Explain why it is necessary to calculate the radius of the circle before you can construct the diagram.

2 Calculate the radius of the circle, and check that your answer agrees with that on the tablet.

Babylonian students also used Pythagoras's result to solve problems about ladders leaning against walls. Here is an example from an Old Babylonian text.

> A ladder of length 0; 30 stands against a wall. How far will the lower end move out from the wall if the upper end slips down a distance of 0; 6 units?

A later, modified, version of this problem can be found on the right-hand side of the New Babylonian tablet whose transcription is shown in Figure 2.14.

Figure 2.14

Activity 2.13   Applying Pythagoras's theorem

Here is a translation of Figure 2.14. The lines refer to lines of the right-hand column of the tablet.

Line 17   A bar set up against a wall. If the top slips down 3 yards,

Line 18   the lower end slides away 9 yards. What is the bar, what the wall? The units are not known. For example, 3 times 3 is 9. 9 times 9 is 1,21. To 9 you will add 1,21.

Line 19   1,30 times 0; 30. 45 is it. The reciprocal of 3 is 0; 20. 0; 20 times 45,

Line 20   15 is the bar. What is the wall? 15 times 15 is 3,45. 9 times 9 is

Line 21   1,21. 1,21 you subtract from 3,45 and it gives 2,24.

Line 22   What times what we have to take to get 2,24?

Line 23   12 times 12 is 2,24. 12 is the wall.

1   Assuming that the bar was vertical before it slipped, calculate the length of the bar, and write a paragraph explaining the Babylonian algorithm.

2   Explain how the second part of the algorithm shows the use of Pythagoras's theorem.

Reflecting on Chapter 2

What you should know
- how the Babylonians represented their numbers
- how to read Babylonian numbers and to calculate with them
- the geometric algorithm for finding square roots
- examples of algorithms for solving quadratic equations of differing forms.

Preparing for your next review
- Reflect on the 'What you should know' list for the chapter. Be ready for a discussion on any of the points.
- Answer the following check questions.

1   Translate this problem and solution into modern notation.

> I have added up seven times the side of my square and eleven times the area: 6; 15. You write down 7 and 11. You multiply 6; 15 by 11: 1,8; 45. You break off half of 7. 3; 30 and 3; 30 you multiply. 12; 15 you add to 1,8; 45. Result 1,21. This is the square of 9. You subtract 3; 30, which you multiplied, from 9. Result 5; 30. The reciprocal of 11 cannot be found. By what must I multiply 11 to obtain 5; 30? 0; 30, the side of the square is 0; 30.

2   Translate this problem and solution into modern notation.

> The surfaces of my two square figures I have taken together: 21; 15. The side of one is a seventh less than the other. You write down 7 and 6. 7 and 7 you multiply: 49. 6 and 6 you multiply. 36 and 49 you add: 1, 25. The reciprocal of 1, 25 cannot be found. By what must I multiply 1, 25 to give me 21; 15? 0; 15. 0; 30 the side. 0; 30 to 7 you raise: 3; 30 the first side. 0; 30 to 6 you raise: 3 the second side.

3   Write down in modern algebraic form the solutions x and y to the simultaneous equations $x + y = a$ and $xy = b$ represented by the Babylonians.

4   Write a summary of not more than one page of the algebra and geometry of the Old Babylonian period. Use any sources available to you.

Practice exercises for this chapter are on page 151.

The Greeks

Introduction

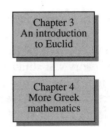

Chapter 3
An introduction
to Euclid

Chapter 4
More Greek
mathematics

Ancient Greece covered an area much greater than modern Greece. It included most of the Mediterranean, and certainly included Sicily, for example, which is where Archimedes lived.

The Greeks had a classification of mathematics into four parts: arithmetic, (numbers at rest), geometry, (magnitudes at rest), music (numbers in motion) and astronomy (magnitudes in motion). In this unit you will consider only the arithmetic and the geometry.

The mathematics of the Greeks developed during the thousand-year period from the 5th century BC to the 5th century AD.

Chapter 3 deals exclusively with Greek geometry, and the very strict rules to which it had to conform. This geometry, and the text book written by Euclid, were to have an enormous effect on mathematical and geometrical thinking, an influence which continues to this day.

Chapter 4 is more of a miscellany, dealing with some other geometric concepts, and with a number of other ideas involving numbers, square roots, algebra and area.

This unit is designed to take about 10 hours of your learning time. About half of this time will be outside the classroom.

Work through the chapters in sequence.

There are summaries and further practice exercises in Chapter 12.

Mathematical knowledge assumed

- some knowledge of methods of proof will be useful.

3 An introduction to Euclid

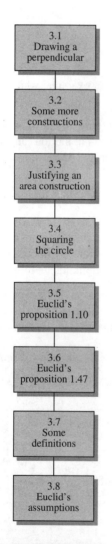

3.1
Drawing a
perpendicular

3.2
Some more
constructions

3.3
Justifying an
area construction

3.4
Squaring
the circle

3.5
Euclid's
proposition 1.10

3.6
Euclid's
proposition 1.47

3.7
Some
definitions

3.8
Euclid's
assumptions

These rules will be called
Euclid's rules. Notice that
they allow neither a set-
square nor a protractor.

In this chapter you will study some of the geometric ideas developed by the ancient Greeks between the 5th century BC and the 5th century AD.

The chapter begins with some construction problems which were characteristic of Greek geometry. It goes on to give some examples of how the Greeks thought about area, and uses some of the construction techniques introduced in the first section to draw squares equal in area to other shapes. Greek geometry was stimulated by attempts to solve three hard problems, which, as we now know, cannot be solved by using the methods to which they restricted themselves. The final two sections are about Euclid and his methods of proof. Activities 3.1 and 3.2 introduce you to some of the Greeks' construction methods, which are taken further in Activity 3.3. Activity 3.4 is about one of the classical Greek problems. Activities 3.5 to 3.8 are about Euclid's deductive methods for proof.

> You will need a straight edge and compasses for Activities 3.1 to 3.3.
>
> All the activities are suitable for small group working, but you should try to work in a group particularly for Activity 3.2.
>
> Work the activities in sequence.

Constructions with a straight edge and compasses

Most of the problems in Greek geometry arose from the philosophical constraint to use only the classical tools for geometrical constructions, a straight edge and a pair of compasses. The straight edge was a ruler with no scale on it, so, although it could be used for drawing a straight line joining two points, it could not be used for measuring distances. The Greek compasses folded together as soon as they were removed from the paper, so they could only be used to draw whole circles or arcs of circles with a given centre and a given radius.

The geometric construction rules were summarised by Euclid in 300 BC as follows.

- You can connect any two given points by a straight line, using a straight edge, and extend the line as far as you want in both directions.
- You can draw a circle with any point as centre and any line segment as radius.
- You can use points of intersection of these figures further on in the construction.

Activities 3.1 and 3.2 contain examples of some Greek geometry problems.

Activity 3.1   *Drawing a perpendicular*

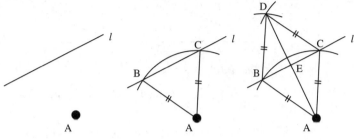

Figure 3.1a   Figure 3.1b   Figure 3.1c

Figure 3.1 shows an example of the use of Euclid's rules to draw a line perpendicular to the line l through the point A.

1   Use the following instructions to make your own copy of Figure 3.1c.
- Take a point B on l and draw the circle with centre A and radius AB.
- This circle intersects l in a second point C.
- Draw circles of radius AB with centres B and C, and mark the point D where these circles intersect.
- Make the triangle BDC, a copy of the isosceles triangle BAC on the other side of l.
- Draw the line AD, and mark the point E where it intersects l.

2   Give reasons why AD is perpendicular to l.

3   Use your diagram so far to draw the circle centre A, which has l as a tangent.

4   Explain why, in Greek mathematics, you would not be allowed to draw the circle in question **3** directly from Figure 3.1a.

Activity 3.2   *Some more constructions*

If you can, work in a group for this activity. Discuss your answers to question 4 in your group.

Two regularly used constructions are the perpendicular bisector of a line-segment, and the bisector of an angle.

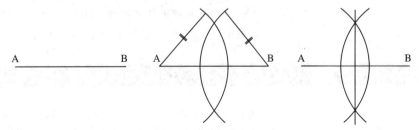

Figure 3.2

1 **a**   Use the progressive drawings in Figure 3.2 to write a list of instructions, similar to those in question **1** of Activity 3.1, for constructing the perpendicular bisector of the line AB.

b In what respect is this construction similar to that of question **1** of Activity 3.1?

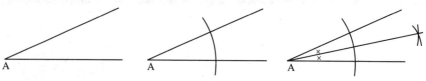

Figure 3.3

2 Similarly, use the drawings in Figure 3.3 to explain how to construct the bisector of the angle at A.

3 What properties of triangles or quadrilaterals do you need to use in explaining why these constructions are correct?

4 Draw any triangle ABC. Construct, using Euclid's rules,
a the bisector of angle A
b the median from angle B
c the line from angle C, perpendicular to AB.

Measuring area – quadrature

An important idea in Greek geometry is that of quadrature; that is, measuring area. Today, if you want to indicate the size of a geometrical object such as the length of a line segment or the area of a polygon, you measure it. To find the area of a plane shape, you choose a unit area and find how many of these units are contained by your shape. When you say that a circle of radius 2 units has an area of about 12.6 square units, you mean that the circle can be approximately paved with 12 unit squares – many of them cut into very small pieces! – plus 6 tenths of such a unit area. Length, area and volume are measured using real numbers, and the area of a geometric figure is described by the appropriate number.

In Greek times, and until the 17th century, mathematicians thought about area and other questions of size differently. To find an area, instead of trying to determine the number of unit areas contained in the given shape, they aimed to transform the shape into a square by geometrical constructions using Euclid's rules. The solution to an area problem was therefore a geometrical construction to produce a square with the same area as the original shape. This construction was called a **quadrature** of the shape. If two shapes had equal areas, they were said to be 'equal'.

Activity 3.3 guides you through the justification of the quadrature of a pentagon.

Activity 3.3   *Justifying an area construction*

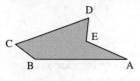

Figure 3.4

You are given the pentagon ABCDE in Figure 3.4, and you have to construct a square of equal area; that is, you have to square the pentagon.

Step 1 is to transform the pentagon into a quadrilateral that has equal area.

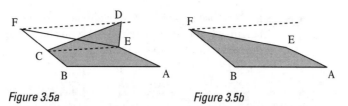

Figure 3.5a　　　　　　　　　*Figure 3.5b*

1 a Study Figure 3.5. What construction steps have been carried out?

b Why is the quadrilateral ABFE equal to the pentagon ABCDE?

Step 2 is to transform ABFE into an equal triangle, AGE, using similar constructions to those used in Step 1.

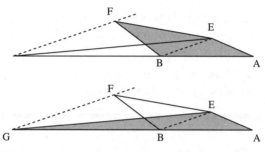

Figure 3.6

Step 3 is to transform triangle AGE into an equal rectangle, GHIJ.

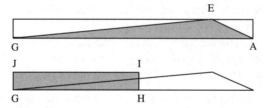

Figure 3.7

c How do you construct GHIJ from AGE?

d Why is the rectangle GHIJ equal to the triangle AGE?

The final part of the problem is to transform a rectangle into an equal square. One of the solutions described by Euclid is based on the idea of the **mean proportional**. If a and b are two line segments, the mean proportional is the line segment x such that $a:x = x:b$.

e Explain why, if a and b are the sides of the rectangle, the mean proportional, x, is the side of the required square.

You can re-formulate this problem as follows: given line segments of lengths a and b, construct a line segment of length x such that $a:x = x:b$.

For this construction Euclid used the fact that the length of the perpendicular in a right-angled triangle is the mean proportional of the two sides adjacent to the right angle.

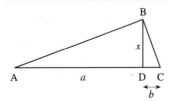

Figure 3.8

f Study Figure 3.8 and explain why, if the perpendicular x divides the hypotenuse in segments a and b, then $a:x = x:b$.

g Now use Figure 3.9 to explain how to construct the square HMLK with sides of length x from the rectangle GHIJ with sides of lengths a and b. Give reasons why the square is equal to the rectangle.

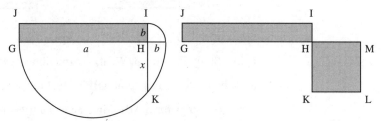

Figure 3.9a   *Figure 3.9b*

In the quadrature of the pentagon, no use was made of numbers. There was no need to know the number of unit lengths contained in a line segment. All that was needed was the construction steps using Euclid's rules.

2 Draw an arbitrary quadrilateral and use similar methods to the steps of question **1** to find a square equal in area. Give reasons for each step you make.

Three classical geometry problems

In the geometry of ancient Greece were three problems destined to have a lasting impact on the development of mathematics in Greece itself and in western Europe as a whole. These problems were

● squaring a circle: constructing a square with the same area as a given circle
● doubling a cube: constructing a cube with twice the volume of a given cube
● trisecting an angle: dividing a given angle into three equal parts.

The fact that these unsolved problems existed gave enormous energy to the practice of mathematics. By the end of the 5th century BC, the problem of squaring a circle was already so well known that Aristophanes, a Greek dramatist of the period 448–385 BC, referred to it in one of his comedies, *The Birds*.

Activity 3.4   Squaring the circle

There have been many attempts at solving the problem of squaring the circle, but nobody has been able to solve the problem using Euclid's rules.

The following sources contain discussions of two attempts to square the circle: the first by Antiphon, and the second by Bryson. The methods used by Antiphon and Bryson, and the essence of their thinking, are demonstrated in Figure 3.10.

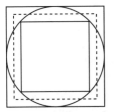

Figure 3.10a

The method of Bryson

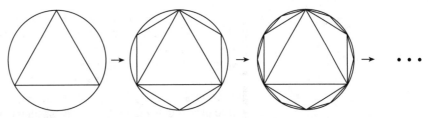

Figure 3.10b

The method of Antiphon

Study Figures 3.10a and 3.10b, read the following quotations and then answer the questions that follow.

From *Physics*, by Aristotle.

> The exponent of any science is not called upon to solve every kind of difficulty that may be raised, but only such as arise through false deductions from the principles of the science: with others than these he need not concern himself. For example, it is for the geometer to expose the quadrature by means of segments, but it is not the business of the geometer to refute the argument of Antiphon.

From a commentary by Themistius on Aristotle's *Physics*.

> For such false arguments as preserve the geometrical hypotheses are to be refuted by geometry, but such as conflict with them are to be left alone. Examples are given by two men who tried to square the circle, Hippocrates of Chios and Antiphon. The attempt of Hippocrates is to be refuted. For while preserving the principles, he commits a paralogism by squaring only that lune which is described about the side of the square inscribed in the circle, though including every lune that can be squared in the proof. But the geometer could have nothing to say against Antiphon, who inscribed an equilateral triangle in the circle, and on each of the sides set up another triangle, an isosceles triangle with its vertex on the circumference of the circle, and continued in this process, thinking that at some time he would make the side of the last triangle, although a straight line, coincide with the circumference.

From *Sofistical refutations* by Aristotle.

> The method by which Bryson tried to square the circle, were it ever so much squared thereby, is yet made sophistical by the fact that it has no relation to the matter in hand ... . The squaring of the circle by means of lunules is not eristic, but the quadrature of Bryson is eristic; the reasoning used in the former cannot be applied to any subject other than geometry alone, whereas Bryson's argument is directed to the mass of people who do not know what is possible and what is impossible in each department, for it will fit any. And the same is true of Antiphon's quadrature.

From Alexander's commentary on Aristotle.

> But Bryson's quadrature of the circle is a piece of captious sophistry for it does not proceed on principles proper to geometry but on principles of more general application. He circumscribes a square about a circle,

inscribes another within the circle, and constructs a third square between the first two. He then says that the circle between the two squares [i.e., the inscribed and the circumscribed] and also the intermediate square are both smaller than the outer square and larger than the inner, and that things larger and smaller than the same things, respectively, are equal. Therefore, he says, the circle has been squared. But to proceed in this way is to proceed on general and false assumptions, general because these assumptions might be generally applicable to numbers, times, spaces, and other fields, but false because both 8 and 9 are smaller than 10 and larger than 7 respectively, but not equal to each other.

1   What was Antiphon's solution to the problem? What was Bryson's solution? Write a summary of each argument in your own words.

2   Why did Aristotle scorn these solutions?

3   Do you think either Antiphon or Bryson had really solved the problem? If not, what are the faults in the arguments?

Modern mathematical methods have since proved that each of the three classical construction problems listed at the beginning of this section is impossible, but generations of Greek mathematicians tried to find solutions. In the course of their searches, they drew many curves and explored their properties.

To solve these problems the rules imposed by Euclid had to be relaxed. You will see one of the ways in which they were relaxed in the work of Archimedes in Chapter 4.

Some propositions from Euclid

Euclid's *Elements* consists of a number of separate books. Activities 3.5 and 3.6 will show you how Euclid's arguments are built up. Each book consists of a series of statements, called propositions. Later propositions are proved from the earlier ones. The proof of each proposition is rather like an algorithm, which can then be used in the algorithm for later propositions.

Activity 3.5   Euclid's proposition 1.10

Read through the following text of proposition 1.10 from *Book I* of Euclid's *Elements*. The numbers in the square brackets refer to propositions proved in *Book I*.

1.10   To bisect a given finite straight line.

Let AB be the given finite straight line.
Let the equilateral triangle ABC be constructed on it,   [1.1]
and let the angle ACB be bisected by the straight line CD: [1.9]
I say that the straight line AB has been bisected at the point D.

For, since AC is equal to CB, and CD is common, the two sides AC, CD are equal to the two sides BC, CD respectively;
and the angle ACD is equal to the angle BCD;

> therefore the base AD is equal to the base BD. [1.4]

> Therefore, the given finite straight line AB has been bisected at D.
>   Q.E.F.

1 Follow the construction in the first paragraph of the proof of proposition 1.10.

2 Proposition 1.1 is: 'On a given straight line to construct an equilateral triangle.' Explain how to use Euclid's rules to do this.

3 Write in your own words the propositions 1.9 and 1.4 referred to in the proof of proposition 1.10.

4 Find out what 'Q.E.F.' stands for and what it means.

Activity 3.6   *Euclid's proposition 1.47*

> **1.47**   In right-angled triangles the square on the side subtending the right angle is equal to the squares on the sides containing the right angle.

1 This proposition is the culmination of the first book of Euclid's *Elements*. What well-known result is it?

2 The next quotation describes the construction used in the proof.

> Let ABC be a right-angled triangle having the angle BAC right:
> I say that the square on BC is equal to the squares on BA, AC.

> For let there be described on BC the square BDEC, and on BA,
> AC the squares AGFB, AHKC;   [1.46]
> through A let AL be drawn parallel to either BD or CE, and let AD, FC be joined.

Follow the construction above to draw the diagram which it describes. Note that proposition 1.46 is the construction of a square with a given straight line as one side.

3 Study the diagram from question **2** carefully, and use it, together with Euclid's proof which follows, to write in your own words a proof of proposition 1.46.

Here is the text of Euclid's proof.

> Then, since each of the angles BAC, BAG is right, it follows that with a straight line BA, and at the point A on it, the two straight lines AC, AG not lying on the same side make the adjacent angles equal to two right angles; therefore CA is in a straight line with AG.   [1.14]

> For the same reason BA is also in a straight line with AH.

> And, since the angle DBC is equal to the angle FBA: for each is right: let the angle ABC be added to each;
> therefore the whole angle DBA is equal to the whole angle FBC.   [C.N.2]

> And, since DB is equal to BC, and FB to BA, the two sides AB, BD are equal to the two sides FB, BC respectively, and the angle ABD is equal to the angle FBC;

The reference C.N.2 is to the second of five 'common notions' or axioms used by Euclid. The second axiom states: 'If equals be added to equals, the wholes are equal.'

> therefore the base AD is equal to the base FC, and the triangle ABD is
> equal to the triangle FBC. [1.4]
> Now the parallelogram BL is double of the triangle ABD, for they have the
> same base BD and are in the same parallels BD, AL. [1.41]
> And the square GB is double of the triangle FBC, for they again have the
> same base FB and are in the same parallels FB, GC. [1.41]
> [But the doubles of equals are equal to one another]
> Therefore the parallelogram BL is also equal to the square GB.
>
> Similarly, if AE, BK be joined, the parallelogram CL can also be proved
> equal to the square HC;
> therefore the whole square BDEC is equal to the two
> squares GB, HC. [C.N.2]
> And the square BDEC is described on BC, and the squares GB,
> HC on BA, AC.
> Therefore the square on the side BC is equal to the squares on the
> sides BA, AC.

Axiomatic structure

The previous two activities involved two examples of the way in which Euclid built
a structure of propositions or theorems, each one depending on the previous ones.
This single deductive system is based upon a set of five initial assumptions, called
postulates, definitions and axioms. Euclid's geometry is like a firm building, based
on a foundation of five pillars (the postulates), in combination with a set of accepted
starting points, the common notions, and using definitions that explain the meaning
of some important ideas.

The first book of the *Elements* opens with a list of 23 definitions in which Euclid
attempts to define precisely the objects and ideas that he is going to use. Each
definition can make use of previously defined terms, but inevitably the first
definitions are difficult to understand.

Activity 3.7   Some definitions

It is difficult to write definitions for a point and a line. Try question **1** briefly before
looking at the answers.

1   Write definitions for 'a point' and 'a line' in your own words.

2   The answers to question **1** give Euclid's definitions of a point and a line. Do you
find them any more helpful than the ones you have written yourself?

3   The next definitions given by Euclid are the following.

> 3   The extremities of a line are points.
>
> 4   A straight line is a line which lies evenly with the points on itself.
>
> 5   A surface is that which has length and breadth only.

6   The extremities of a surface are lines.

7   A plane surface is a surface which lies evenly with the straight lines on itself.

8   A plane angle is the inclination to one another of two lines in a plane which meet one another and do not lie in a straight line.

Criticise definition 7.

4   Read the following definitions, and write in your own words a definition of a circle.

15   A circle is a plane figure contained by one line such that all the straight lines falling upon it from one point among those lying within the figure are equal to one another;

16   And the point is called the centre of the circle.

17   A diameter of the circle is any straight line drawn through the centre and terminated in both directions by the circumference of the circle, and such a straight line also bisects the circle.

5   Write an addition to definition 17 so that it includes a definition of the radius.

Following the definitions, Euclid gives two lists of statements that he assumes to be true in all his subsequent work. Any mathematical work assumes some things to be true, and takes these as **axioms**.

Euclid's axioms come in two separate lists: the five postulates and the five common notions. It is not clear why Euclid made this separation, but another Greek thinker, Aristotle, taught that common notions are truths that underlie all deductive thinking and are convincing in themselves, whereas postulates are less obvious and may apply only to a particular study. Modern mathematicians no longer make this distinction; they call both types axioms.

Activity 3.8   *Euclid's assumptions*

Here are Euclid's axioms.

POSTULATES

Let the following be postulated:

1   To draw a straight line from any point to any point.

2   To produce a finite straight line continuously in a straight line.

3   To describe a circle with any centre and distance.

4   That all right angles are equal.

5   That, if a straight line falling on two straight lines make the interior angles on the same side less than two right angles, the two straight lines, if produced indefinitely, meet on that side on which are the angles less than the two right angles.

COMMON NOTIONS

1   Things which are equal to the same thing are also equal to one another.
2   If equals be added to equals, the wholes are equal.
3   If equals be subtracted from equals, the remainders are equal.
4   Things which coincide with one another are equal to one another.
5   The whole is greater than the part.

> Interpret postulate 3 in the very limited sense implied by Euclid's rules.

1   The first two postulates assume that it is always possible to draw a straight line between any two given points, and that such a line can always be extended. Write your own explanations of what is meant by each of postulates 3 and 4.

2   You can re-write common notion 1 in modern mathematical notation as

$$a = b \text{ and } b = c$$
$$\Rightarrow a = c$$

Do the same for common notions 2 and 3.

3   Postulate 5 is often known as 'the parallel postulate'. Why do you think this is?

Aristotle also taught that 'other things being equal, that proof is the better which proceeds from the fewer postulates'. This principle is still important today, and attempts by mathematicians to reduce the number of postulates required by Euclid's *Elements* have stimulated much mathematical activity since his time.

The parallel postulate in particular has caused much discussion over the centuries. Many mathematicians since Euclid have believed that it should be possible to prove it from the other axioms and postulates. For example, the Arab mathematician, al Haytham, set out to prove the parallel postulate using only the first 28 propositions from the *Elements*. These are the propositions for the proof of which the parallel postulate is not required. His method of proof was later criticised by Omar Khayyam in the 12th century. Another great mathematician, Nicholas Lobachevsky, wrote in 1823 about the parallel postulate that 'no rigorous proof of this has ever been discovered'. Three years later he had begun to produce a new type of geometry, a non-Euclidean geometry, which was built specifically on an assumption in opposition to Euclid's parallel postulate. Work on this new geometry was not accepted at first, and it was not until about 30 years later that Riemann, another mathematician, saw the work thoroughly integrated into mainstream mathematics.

Reflecting on Chapter 3

What you should know

- that Euclid established an axiomatic method by his approach to geometry, and some of the details of his methods
- the Greeks' rules for geometric constructions
- how to carry out the constructions for drawing a perpendicular, for bisecting an angle, for bisecting a line, and for constructing a perpendicular bisector of a line

- the meaning of 'quadrature'
- how to draw a square equal in area to a given polygon
- the three classical geometric problems.

Preparing for your next review

- Reflect on the 'What you should know' list for this chapter. Be ready for a discussion on any of the points.
- Answer the following check questions.

1 Use the methods of this chapter to construct a square equal in area to an equilateral triangle.

2 Use Euclid's rules, and Euclid's proposition 1.1, to construct a regular hexagon.

Practice exercises for this chapter are on page 152.

4 More Greek mathematics

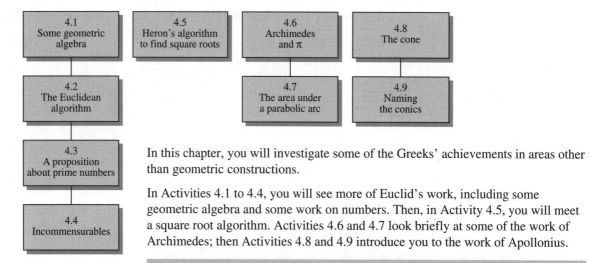

In this chapter, you will investigate some of the Greeks' achievements in areas other than geometric constructions.

In Activities 4.1 to 4.4, you will see more of Euclid's work, including some geometric algebra and some work on numbers. Then, in Activity 4.5, you will meet a square root algorithm. Activities 4.6 and 4.7 look briefly at some of the work of Archimedes; then Activities 4.8 and 4.9 introduce you to the work of Apollonius.

All the activities are suitable for small group working.

You should work Activities 4.1 to 4.4 in sequence.

Book II of Euclid's *Elements* is short, containing only 14 propositions, each of which is concerned with geometric algebra. Modern mathematics at secondary school level makes extensive use of algebraic notation; magnitudes are represented by letters that are understood to be numbers, and arithmetic operations can be applied to these under the usual rules of algebra. The Greek mathematicians of Euclid's time used line segments to represent magnitudes, and they operated on these using the theorems and axioms of geometry.

Activity 4.1   Some geometric algebra

1   The following is proposition 1 from *Book II* of the *Elements*.

> If there be two straight lines, and one of them be cut into any number of segments whatever, the rectangle contained by the two straight lines is equal to the rectangles contained by the uncut straight line and each of the segments.

Draw a diagram to illustrate this proposition in the case where the cut straight line is cut into three line segments.

2 With reference to your diagram, what is the proposition saying?

3 Write an algebraic statement that is equivalent to the proposition.

4 Proposition 4 from *Book II* states:

> If a straight line be cut at random, the square on the whole is equal to the squares on the segments and twice the rectangle contained by the segments.

Illustrate this proposition and write down an equivalent algebraic statement.

Activity 4.2   The Euclidean algorithm

Book VII of the *Elements* contains 39 propositions. The first 19 deal with ratios and proportions of integers. Proposition 1 introduces the procedure now known as Euclid's algorithm, which plays an important part in the theory of numbers.

> Two unequal numbers being set out, and the less being continually subtracted from the greater, if the number which is left never measures the one before it until a unit is left, the original numbers will be prime to one another.

A number which when subtracted several times from another gives a zero remainder is said to **measure it exactly**.

1 Study the proposition above, and use it to write an algorithm which takes two positive whole numbers as input, and, as output, states whether or not the numbers are prime to one another; that is, whether or not they are co-prime.

2 Program your graphics calculator to carry out the Euclidean algorithm, and check that it reports correctly whether or not the two input numbers are co-prime.

3 Proposition 2 of *Book VII* makes use of the Euclidean algorithm.

> Given two numbers not prime to one another, to find their greatest common measure.

Write down in your own words what 'greatest common measure' means.

4 Experiment with your graphics calculator program to find out how the Euclidean algorithm can be used to find the greatest common measure.

Activity 4.3   A proposition about prime numbers

Later books in the *Elements*, *Books VII* to *IX*, deal with the theory of numbers. In *Book IX*, proposition 20 is particularly well-known:

> Prime numbers are more than any assigned multitude of prime numbers.
>
> Let A, B, C be the assigned prime numbers: I say that there are more prime numbers than A, B, C.

For let the least number measured by A, B, C be taken, and let it be DE; let the unit DF be added to DE.

Then EF is either prime or not.

First let it be prime; then the prime numbers A, B, C, EF have been found which are more than A, B, C.

Next let EF not be prime; therefore it is measured by some prime number.   [VII.31]

Let it be measured by the prime number G.

I say that G is not the same with any of the prime numbers A, B, C.

For, if possible, let it be so.

Now A, B, C measure DE; therefore G will also measure DE.

But it also measures EF.

Therefore G, being a number, will measure the remainder, the unit DF; which is absurd.

Therefore G is not the same with any of the numbers A, B, C.

And by hypothesis it is prime.

Therefore the prime numbers A, B, C, G have been found which are more than the assigned multitude of A, B, C.   Q.E.D.

> Q.E.D. stands for 'quod erat demonstrandum', which translates into 'which was to be demonstrated', but everyone says 'quite enough done'!

This proposition states that the number of prime numbers is infinite. Euclid uses an indirect method of proof, assuming that there is a finite number of prime numbers, and showing that this assumption leads to a contradiction.

1   Construct a proof in modern notation, using Euclid's method.

Incommensurability

The Greek scholar Pythagoras taught that 'all things are numbers'. He and his followers, the Pythagoreans, had discovered a relationship between whole numbers and music, and they held mystical beliefs about the significance of numbers. They believed that all possible numbers could be expressed as ratios between whole numbers, and that anything, whether geometry or human affairs, could be explained in terms of whole numbers represented by dots or finite indivisible elements of some sort. The sides and diagonals of certain rectangles can be represented by equally spaced dots as shown in the first two diagrams in Figure 4.1, since $3^2 + 4^2 = 5^2$ and $5^2 + 12^2 = 13^2$. In such cases you can express the ratio between the diagonal and the side as the ratio of a pair of whole numbers. Then the side and the diagonal are called **commensurable**. If you cannot express the ratio in this way, as shown in the third diagram in Figure 4.1, then the numbers are **incommensurable**.

The discovery of the famous proposition about right-angled triangles lent support to the Pythagorean view, but it led to the discovery of incommensurable quantities. This was a shock to the Pythagoreans, and had a significant effect on the development of mathematics in ancient Greece. A number system consisting only of whole numbers and the ratios between them was not enough to represent relationships between quantities such as line segments. As a result, the mathematics of the ancient Greeks moved away from the arithmetic emphasis of the Pythagoreans, and became almost entirely geometric.

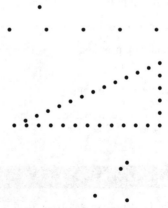

Figure 4.1

Activity 4.4   Incommensurables

Here is definition 1 from *Book X* of Euclid's *Elements*.

> Those magnitudes are said to be commensurable which are measured by the same measure, and those incommensurable which cannot have any common measure.

A

n units

B   *m* units   C

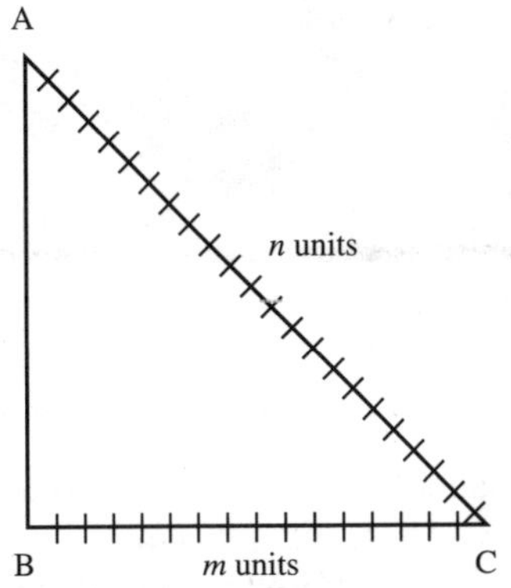

Figure 4.2

This proof is an early example of proof by contradiction. The proof of the irrationality of $\sqrt{2}$ appeared in the *Investigating and proving* unit in Book 4.

1 Explain Euclid's definition in your own words.

2 The proposition given below was at one time recorded as proposition 117 in *Book X* of Euclid's *Elements*, but it has now been shown that it was not part of Euclid's original text. Even so, it certainly originates in ancient Greece.

> Proposition: the diagonal of any square is incommensurable with the side.
>
> Suppose the diagonal AC, and the side AB, of a square are commensurable. Then there exists a unit length in terms of which AC is *n* units and AB is *m* units, and *n* and *m* are the smallest possible. This means that the numbers *n* and *m* have no divisors in common (other than the unit). Since the square on the diagonal AC is twice the square on the side AB, we have $n^2 = 2m^2$.
>
> It follows that n^2 is an *even* number (since it is *twice* some number) — so *n* is an even number; and so *m* *must* be an *odd* number (because *n* and *m* have no divisors in common). Since *n* is even, $n = 2p$ (say), and its square being equal to $2m^2$ gives us $(2p)^2 = 2m^2$; so $4p^2 = 2m^2$; so $2p^2 = m^2$. Thus m^2, and so *m*, now turns out to be an *even* number. But we already know that *m* must be an odd number. So we have shown that *m* is both an even number and an odd number, which is absurd.

Use this method to write in your own words a proof that $\sqrt{3}$ is irrational.

In spite of the fact that the main emphasis of Greek mathematics was not on the arithmetic evaluation of quantities, some Greek mathematicians developed sophisticated methods of evaluation. Heron of Alexandria, who lived in the 1st century AD, wrote a book, *Metrica*, concerned mainly with numerical examples of practical methods for calculating lengths, areas and volumes. Unlike Euclid's *Elements*, Heron's work is more of a set of rules, not all of which were found to be correct in the light of subsequent knowledge, for solving certain types of problems. Activity 4.5 gives, as an example, Heron's method for calculating square roots. The method models a square as a rectangle with sides that gradually approach the same length.

Activity 4.5   Heron's algorithm to find square roots

The area of the square is A. To find the side
- choose any length, x, for the side of a rectangle,
- calculate the other side of the rectangle as $\dfrac{A}{x}$, since $x \times \dfrac{A}{x} = A$ as required
- find the average of x and $\dfrac{A}{x}$

- treat this average as a new estimate for the length of a side and repeat the process until the estimate of $\sqrt{A}$ is sufficiently accurate.

1 a Write in algorithmic notation an algorithm to carry out this iteration to evaluate $\sqrt{N}$, until successive approximations differ by less than 10^{-6}.

b Use your algorithm on your graphics calculator to evaluate $\sqrt{5}$.

2 Where have you seen this algorithm before?

3 Modify Heron's method to find a cube root.

Archimedes

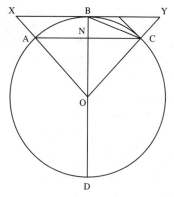

Figure 4.3

Archimedes has been described as the greatest mathematician of antiquity. He lived from 287 to 212 BC, in Syracuse in Sicily, as a royal adviser. He was killed by Roman soldiers when the Romans took Syracuse, after he had offered his help to the king in defence of the city. He was a man of considerable technical skill, 'an ingenious inventor and a great theoretician at the same time'. Accounts of the life of Archimedes agree that he placed much greater value on his theoretical achievements than on the mechanical devices that he made. Archimedes took the ideas of Bryson and Antiphon (see Activity 3.4) a great deal further in an attempt to calculate π.

Archimedes started from hexagons inside and outside the circle, and calculated their semi-perimeters, so that the value of π lay between these values. He then doubled the number of sides, and continued this process. Figure 4.3 shows triangles OAC and OXY after n such doublings – they are not drawn to scale because 6×2^n of these triangles together form a regular 6×2^n-sided polygon.

In the description of Archimedes's method which follows in Activity 4.6, you must remember that he did not have trigonometry at his disposal.

Activity 4.6   Archimedes and π

Start from the situation shown in Figure 4.4, where the radius of the circle is 1 unit.

Figure 4.4

1 Let S_0 be the length outlined inside the circle and T_0 be the length outside.

Show, without trigonometry, that $S_0 = 3$ and $T_0 = \dfrac{6}{\sqrt{3}}$. Hence $3 < \pi < \dfrac{6}{\sqrt{3}}$.

2 Figure 4.5 is the same as Figure 4.3, with some extra lines and labels added.

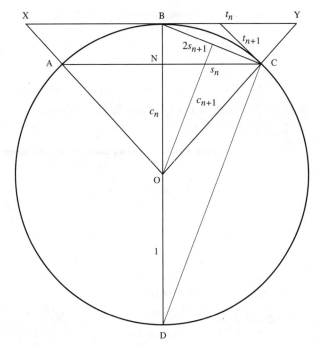

Figure 4.5

Notice that Figure 4.5 is not drawn to scale.

In Figure 4.5, $NC = s_n$ and the chord AC with length $2s_n$ is at a distance c_n from the centre of the circle. In the triangle OBC, the chord BC from the next iteration has length $2s_{n+1}$, and its distance from the centre is c_{n+1}. The tangent at B has length $2t_n$, while the tangent at C, from the next iteration, has total length $2t_{n+1}$.

a Show that $CD = 2c_{n+1}$.

b By calculating the area of the triangle BCD in two different ways, show that $s_n = 2s_{n+1}c_{n+1}$.

c Use the similar triangles CDN and BDC to show that $\dfrac{2c_{n+1}}{1+c_n} = \dfrac{2}{2c_{n+1}}$, and hence that $c_{n+1} = \sqrt{\dfrac{1+c_n}{2}}$.

d Use similar triangles to show that $t_n = \dfrac{s_n}{c_n}$.

Let S_n be half the perimeter of the inside 6×2^n-sided polygon after n iterations, and T_n be the corresponding semi-perimeter of the corresponding outside polygon.

3 a Show that $S_n = 3 \times 2^{n+1} s_n$ and $T_n = 3 \times 2^{n+1} t_n$.

b Use the results of part **a** and of question **2** to show that

$$S_{n+1} = 3 \times 2^{n+2} s_{n+1} = \frac{3 \times 2^{n+1} s_n}{c_{n+1}} = \frac{S_n}{c_{n+1}} \quad \text{and}$$

$$T_{n+1} = 3 \times 2^{n+2} \times t_{n+1} = \frac{3 \times 2^{n+2} s_{n+1}}{c_{n+1}} = \frac{S_{n+1}}{c_{n+1}}.$$

4 Use the three recurrence relations

$$c_{n+1} = \sqrt{\frac{1+c_n}{2}}, \qquad S_{n+1} = \frac{S_n}{c_{n+1}} \qquad \text{and} \qquad T_{n+1} = \frac{S_{n+1}}{c_{n+1}}$$

together with the starting conditions from question **1**, and the correct value of c_0, to write an algorithm for your calculator to calculate and display values of S_n and T_n until they differ by less than 10^{-10}. You will need to put a pause statement into the algorithm at each iteration.

5   What would be the modern way of establishing these recurrence relations?

Archimedes actually carried out four iterations starting from the regular hexagon in Figure 4.4. Thus he reached the approximation $3.1403\ldots < \pi < 3.1427\ldots$.

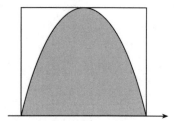
Figure 4.6

The process which Archimedes used to find the value of π is an example of the process of exhaustion. Archimedes showed that the only number which is greater than the areas of all the inside regular polygons, and, at the same time, is less than the areas of all the outside regular polygons is $3.141\,59\ldots$.

A geometric argument shows that the circle has the property that its area, π, is greater than the areas of all the inside regular polygons, and less than the areas of all the outside regular polygons. The process of exhaustion is then used to show that the area of the circle, π, is $3.141\,59\ldots$. All other possibilities have been exhausted.

Some of the ideas of modern integration can be traced back to the method which Archimedes used to calculate his approximations to π.

Archimedes also showed that the area under a symmetrical parabolic arc is $\frac{2}{3}$ of the area of the rectangle which surrounds it, as shown in Figure 4.6.

Archimedes's method, however, is not general. It relies on a geometric property of the parabola, which it is helpful to prove as a first step.

Activity 4.7   The area under a parabolic arc

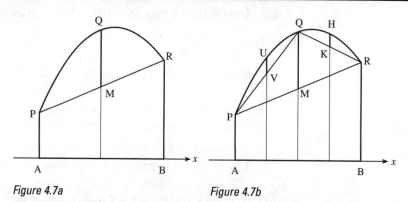

Figure 4.7a   Figure 4.7b

In Figure 4.7a, M is the mid-point of PR, and the length QM is the distance between the parabola and the chord PR. In Figure 4.7b, V and K are the mid-points of PQ and QR. The crucial step in Archimedes's proof is that, for any parabola,
$UV = HK = \frac{1}{4}QM$.

Suppose that the parabola has the equation $f(x) = px^2 + qx + r$, and that the points A and B have x-coordinates a and b respectively. Let $b - a = h$.

1 a Prove that $QM = -\frac{1}{4}ph^2$.

b Why is there a negative sign in this expression?

c Without doing further detailed calculations, show how you can say immediately that $UV = HK = -\frac{1}{4}p\left(\frac{1}{2}h\right)^2 = -\frac{1}{16}ph^2$.

2 a Now go back to the symmetric parabola shown in Figure 4.8. Show that a suitable equation for it is $f(x) = px^2 + r$.

b Show that the area of triangle PRQ is ar. Apply the result of question **1c** to show that the area of each of the triangles QHR and PUQ is $\frac{1}{8}ar$, so that the area of the shape PUQHR is $ar + 2 \times \frac{1}{8}ar = ar + \frac{1}{4}ar$.

c Show that, if you continue this process by creating more triangles in the same way, the areas of successive approximations to the area of the parabola are

$$ar$$

$$ar + \tfrac{1}{4}ar$$

$$ar + \tfrac{1}{4}ar + \tfrac{1}{16}ar$$

$$ar + \tfrac{1}{4}ar + \tfrac{1}{16}ar + \tfrac{1}{64}ar$$

$$\vdots$$

d Find the sum of this series, and so find the area under the parabolic arc.

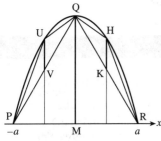

Figure 4.8

Curves

The period of Greek history between about 300 and 200 BC is sometimes described as 'The Golden Age of Greece'. During this time there was a remarkable flowering of the arts and literature, and of mathematics. Three names are principally associated with the development of mathematics in Greece at this time. The work of two of these, Euclid and Archimedes, has been discussed already. The third of these great mathematicians was Apollonius of Perga. Little is known of his life, and many of his writings have been lost. It is thought that he lived from around 262 to 190 BC, and he appears to have studied at Alexandria with the successors of Euclid, before settling at Pergamum where there was an important university. Many of his writings were briefly described by Pappus about 500 years later in a work called the *Collection*. The descriptions given by Pappus and others were extensively used in modern times, by European mathematicians of the 17th century including Viète, van Schooten and Fermat, who devoted considerable effort to reconstructing some of the lost works of Greek mathematicians.

> Other contributions by Viète, van Schooten and Fermat are discussed later in the *Descartes* unit.

As well as being an outstanding mathematician of his day, Apollonius was also known for his work on astronomy. He proposed important ideas about planetary motion, which were later used by Ptolemy (about AD 140) in his book, *Almagest*, summarising the astronomy known to the ancient Greeks.

Apollonius of Perga is best known today for a major treatise on conic sections, the only one of his major works substantially to have survived. This is a collection of eight books, begun while Apollonius was in Alexandria, and continued from Pergamum. The first four books draw together the theoretical work of earlier mathematicians, including Euclid and Archimedes. The later books developed the ideas substantially. This work was highly influential on later mathematicians in

49

Greece and beyond, and as a result of it Apollonius became known as 'The Great Geometer'.

Apollonius was responsible for naming the conic sections: parabola, hyperbola and ellipse. The names refer to area properties of the curves.

Activity 4.8   The cone

In the definitions in *Book I* of *Conics*, Apollonius defines his cone.

> If a straight line infinite in length and passing through a fixed point be made to move around the circumference of a circle which is not in the same plane with the point so as to pass successively through every point of that circumference, the moving straight line will trace out the surface of a double cone.

These sections were described in Activity 5.6 of the *Plane curves* unit in Book 5.

1   Define in your own words the vertex of the cone and the axis of the cone.

2   With the help of diagrams, describe briefly how different sections of the shape described above produce the parabola, the hyperbola, and the ellipse.

Apollonius's definition of the cone was more general than earlier definitions in that the axis is not required to be perpendicular to the base, and the cone extends in both directions. He went on to investigate the different plane sections of the cone.

Activity 4.9   Naming the conics

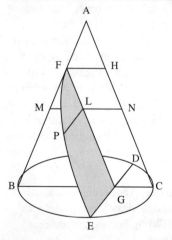

Figure 4.9

Apollonius gave names to the line segments FL and LP, which are usually translated as 'abscissa' and 'ordinate' respectively.

Questions 2 and 3 are more difficult than the other question and are optional.

Consider a cone with vertex A, and a circular base with diameter BC (Figure 4.9). ABC is a vertical plane section of the cone through its axis. Let DE be a chord on the circular base, perpendicular to the diameter BC, intersecting BC at G, and let F be a point on the line AB. Then DFE defines a section of the cone, and FG is called a diameter of the section. Let the line MN be the diameter of a section of the cone parallel to the base, cutting FG at L. P is the point on the circumference of this section where it cuts DFE.

Finally, let FH be the diameter of the section of the cone parallel to the base.

Apollonius considered three cases, depending on the angle that the line FG makes with AC.

1   In the first case FG is parallel to AC.

a   Explain why $LN = FH$.

b   Explain why $LP^2 = LM \times LN$. You will find it helpful to refer back to Activity 3.3, questions **1e** and **1f**.

c   Explain why $LM = FL \times \dfrac{HF}{AH}$.

d   Hence show that for any section MNP parallel to the base, $LP^2 = k \times FL$, where k is a constant for the section DFE.

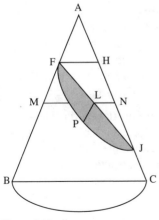

Figure 4.10

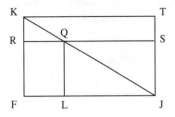

Figure 4.11

This result, in the words of Apollonius is:

> 'if we apply the square LP to a straight line length k, we obtain the abscissa FL', or 'the exact parabola of the square of the ordinate on the line segment k gives the abscissa'.

This is based on a much earlier use of the word 'parabola', possibly introduced by the Pythagoreans. Hence Apollonius named this curve the parabola.

2 In the second case, FL cuts AB and AC at the points F and J (Figure 4.10).

a Use an argument like the one in question **1** to show that $LP^2 = LM \times LN$ is proportional to $LF \times LJ$.

b Suppose that the constant of proportionality is the ratio of some length FK to the length FJ (Figure 4.11).

Show that there is a point Q on KJ such that LP^2 is equal to the rectangle FLQR.

In Apollonius's words, this is to say that 'the square of LP is applied to the line FJ in ellipsis by a rectangle LJSQ, similar to the rectangle FJTK'. Here, the term 'ellipsis' means 'deficiency'. Hence this conic section was named the ellipse.

3 In the third case, FL cuts AB at F, and cuts CA produced at J.

The Greek word 'hyperbola' means 'in excess' or 'a throwing beyond'. Use an argument similar to that in question **2** to explain why Apollonius named the curve in this third case the hyperbola.

The way in which Apollonius worked with the conics, using abscissa and ordinates, suggests that he was very close to developing a system of coordinates. Many of his results can be re-written almost immediately into the language of coordinates, even though no numbers were attached to them. As a result, it has sometimes been claimed that Apollonius had discovered the analytic geometry which was developed some 1800 years later. However, the fact that the Greeks had no algebraic notation for negative numbers was a great obstacle to developing a coordinate system.

Greek mathematicians classified construction problems into three categories. The first class was known as 'plane loci', and consisted of all constructions with straight lines and circles. The second class was called 'solid loci' and contained all conic sections. The name appears to be suggested by the fact that the conic sections were defined initially as loci in a plane satisfying a particular constraint, as they are today, but were described as sections of a three-dimensional solid figure. Even Apollonius derived his sections initially from a cone in three-dimensional space, but he went on to derive from this a fundamental property of the section in a plane, which enabled him to dispense with the cone. The third category of problems was 'linear loci', which included all curves not contained in the first two categories.

Hypatia

Hypatia of Alexandria was born in about AD 370 towards the end of the Greek era. She was the daughter of Theon of Alexandria, known for his commentary on Ptolemy's *Almagest* and his edition of Euclid's *Elements*.

Hypatia's main contribution appears to have been one of continuing the work of her father, writing commentaries on the works of the great Alexandrian mathematicians, Apollonius and Diophantus. She is thought to have written a commentary on an astronomical table of Diophantus, on Diophantus's *Arithmetic,* on work related to Pappus, on Apollonius's work on conics, and to have written her own work about areas and volumes. She became a respected teacher of the Greek classics.

She was invited to be president of the Neoplatonic School at Alexandria, which was at the time one of the last institutions to resist Christianity, and has been described as one of the last pagan, that is, non-Christian, philosophers. She was martyred in AD 415. Possibly because of her martyrdom she occupies an exalted place in the history books. Her writings have been lost, so it is hard to evaluate her work objectively.

The commentaries of others of her time have provided the world with one of the best sources of information about the development of Greek mathematics.

Since it was hard to understand Euclid's *Elements* or Apollonius's *Conics* without help from teachers and commentators, Hypatia's death can be thought of as ending the Greek era.

Reflecting on Chapter 4

What you should know

- the contributions of Euclid, Archimedes, Apollonius and Hypatia
- how to interpret Euclidean theorems of geometric algebra in terms of modern algebra
- what is meant by incommensurability
- how the Greeks proved that some square roots are incommensurable
- why the conic sections are named 'parabola', 'ellipse' and 'hyperbola'.

Preparing for your next review

- Reflect on the 'What you should know' list for this chapter. Be ready for a discussion on any of the points.
- Answer the following check questions.

1 Make brief notes on the lives and contributions of Euclid, Archimedes, Apollonius, and Hypatia.

2 Write in your own words a paragraph about the significance to the Greeks of the discovery of incommensurable quantities.

3 Prepare some examples of each of the three categories of construction problems.

Practice exercises for this chapter are on page 153.

3

The Arabs

Introduction

Chapter 5
Arab mathematics

In this unit, the term 'Arab' refers to a civilisation containing many ethnic, religious and linguistic groups that flourished between the 9th and 15th centuries AD. The Arab civilisation included 'non-Arab' lands, such as present-day Iran, Turkey, Afghanistan and Pakistan, all of which now have distinctive Islamic cultures. The mathematicians whose work is described in this unit all worked in Arabic.

The unit consists of just one chapter which summarises some of the achievements of these 'Arab' mathematicians.

This unit is designed to take about 10 hours of your learning time. About half of this time will be outside the classroom.

There are summaries and further practice exercises in Chapter 12.

Mathematical knowledge assumed

- how to solve quadratic equations.

5 Arab mathematics

You should think of the role of the Arab mathematicians as having two components, which are complementary:

- the translation and subsequent handing on of Greek and Indian mathematics texts, such as Apollonius's *Conics*
- their own original contributions, made by embracing and developing both eastern and western mathematics.

This chapter falls into four parts.

In the first part you learn about the evolution of the Hindu-Arabic number system, and see in Activity 5.1 how the facility with fractions of a whole was used to help implement the Islamic rules of inheritance.

In the second part, Activities 5.2 and 5.3, you will learn how the Arab mathematicians took what was known about trigonometry and developed it further.

In Activities 5.4 to 5.7 you will study some of the variety of problems, originating from Indian and Chinese texts, on which the Arab mathematicians worked.

Then you go on to learn how the Arab mathematicians developed a geometric approach, inspired by Greek geometry, to the algorithmic solution of equations.

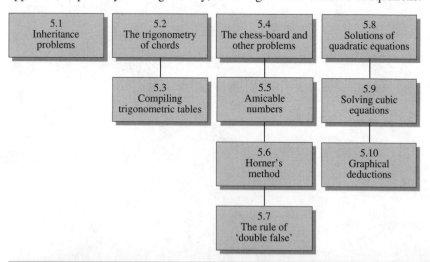

Although the chapter has been designed to be read and worked in sequence, the four parts are independent, and if you prefer you may study them in some other order.

If possible work in a group for Activity 5.8. You may be asked to make a short report based on your work in this activity.

Historical background

The period of Arab intellectual ascendancy can be thought of as beginning with the expansion of Islam in the first half of the 7th century. After a period of over a century of war and expansion a new capital of the Arab civilisation was established at Baghdad, which gradually became a centre of mathematical activity.

A 'House of Wisdom' was established, and mathematical and other scholars from the whole Arab civilisation, as well as India and Greece, were welcomed. The Arabs quickly absorbed the learning of their neighbours.

A multiplicity of influences therefore went into the creation of Arab mathematics.

First, there was Babylonian mathematics and astronomy dating back over 2500 years. This mathematics was passed down either through the medium of a religious sect known as Sabeans, descendants of Babylonian star-worshippers who showed a particular aptitude for astronomy and numerology, or through a number of practical algorithms for solving linear and non-linear equations in one or more unknowns, similar to the old Babylonian methods described in Chapter 2.

Second were the Greek written sources, regarded by many historians of mathematics as of paramount importance in the development of Arab mathematics. Central to the Greek tradition, as it was taken over by the Arab world, were Euclid's geometrical text, *Elements*, and Ptolemy's astronomical text, *Almagest*.

Third, and less certain, were the Indian written sources. The line of influence is not so clear as that of the Greek mathematicians. But, it is clear that the Indians did much work on trigonometry and astronomy, which entered Arab mathematics through the Arabs' interest in astronomy. For example, an astronomical work from India, one of the Siddhantas, was translated into Arabic in about AD 775.

Other knowledge came from a variety of sources. There existed a vast treasure trove of commercial practices, measurement techniques, number reckoning, recreational problems and other examples of 'oral mathematics' which resulted from large groups of merchants, diplomats, missionaries and travellers meeting one another along the caravan routes and the sea routes from China to Cadiz in Spain. The computational procedures or algorithms, arising from this source, often seemed more like recipes than well-thought-out mathematical methods. But a number of Arab mathematicians, inspired by these recipes, proceeded to devise their own techniques for solving problems. In doing so, they sometimes translated their recipes into a more suitable, theoretical form.

Numerals

One of the most far-reaching inventions of Indian mathematics was a positional, or place value, decimal numeral system. With symbols or digits to represent numbers from zero to nine, 0, 1, 2, …, 9, it became possible to represent any integer both uniquely and economically. Indian mathematicians created an arithmetic in which calculations could be carried out by people of average ability. The story of the development and the spread of Indian numerals, often known as the Hindu-Arabic numerals, is fascinating; the role of the Arabs in this development was crucial.

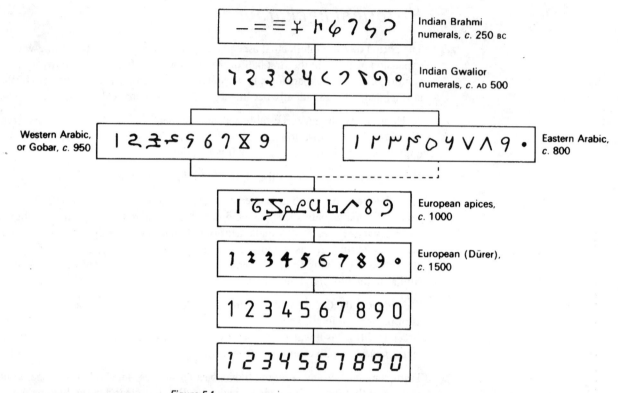

Figure 5.1

You can find out more about the origins, development and the global spread of numerals in, for example, *The crest of the peacock*, by G G Joseph.

You can see from Figure 5.1 how our system of numerals evolved. The development started from certain inscriptions on the Ashoka pillars (stone pillars built by Ashoka, 272–232 BC, one of the most famous emperors of ancient India) and has now reached a time when computers form numbers using pixels.

Both the Babylonians and the Chinese had numeral systems versatile enough to represent and manipulate fractions. But neither had a device, or symbol, for separating integers from fractions. The credit for such a symbol must go to the Arabs. For example, in giving the value of π, the ratio of the circumference of a circle to its diameter, to a higher degree of accuracy than before, as

sah-hah

3   141 592 653 59

where the Arabic word 'sah-hah' above the 3 was in effect the decimal point.

Inheritance problems

Mathematics was recruited to the service of Islamic law. These laws were fairly straightforward; when a woman died her husband received one-quarter of her estate, and the rest was divided among the children in such a way that a son received twice as much as a daughter; when a man died a widow was normally entitled to one-eighth while a mother was entitled to one-sixth. However, if a legacy was left to a stranger, the division became more complicated. The law stated that a stranger could not receive more than one-third of the estate without the permission of the natural heirs. If some of the natural heirs endorsed such a legacy but others did not, those who did must pay, pro-rata, out of their own shares, the amount by which the stranger's legacy exceeded one-third of the estate. In any case, the legacy to the stranger had to be paid before the rest of the estate was shared out among the natural heirs. You can clearly construct problems of varying degrees of complexity which illustrate different aspects of the law. The following example is typical.

> A woman dies leaving a husband, a son and three daughters. She also leaves a bequest consisting of $\frac{1}{8} + \frac{1}{7}$ of her estate to a stranger. Calculate the shares of her estate that go to each of her beneficiaries.

Here is the solution. The stranger receives $\frac{1}{8} + \frac{1}{7} = \frac{15}{56}$ of the estate, leaving $\frac{41}{56}$ to be shared out among the family. The husband receives one-quarter of what remains; that is, $\frac{1}{4}$ of $\frac{41}{56} = \frac{41}{224}$. The son and the three daughters receive their shares in the ratio $2:1:1:1$; that is, the son's share is two-fifths of the estate after the stranger and husband have taken their bequests. So, if the estate is divided into $5 \times 224 = 1120$ equal parts, the shares received by each beneficiary will be:

Stranger: $\frac{15}{56}$ of $1120 = 300$ parts   Husband: $\frac{41}{224}$ of $1120 = 205$ parts
Son: $\frac{2}{5}$ of $(1120 - 505) = 246$ parts   Each daughter: $\frac{1}{5}$ of $(1120 - 505) = 123$ parts.

Activity 5.1   Inheritance problems

Share out the estate for each beneficiary of the following inheritance problems. Think how difficult they would be if you used the Babylonian number system.

1   A man dies leaving four sons and his widow. He bequeaths to a stranger as much as the share of one of the sons less the amount of the share of the widow.

2   A man dies and leaves two sons and a daughter. He also bequeaths to a stranger as much as would be the share of a third son, if he had one.

3   A man dies and leaves a mother, three sons and a daughter. He bequeaths to a stranger as much as the share of one of his sons less the amount of the share of a second daughter, in case one arrived after his death.

4   A man dies and leaves three sons. He bequeaths to a stranger as much as the share of one of his sons, less the share of a daughter, supposing he had one, plus one-third of the remainder of the one-third.

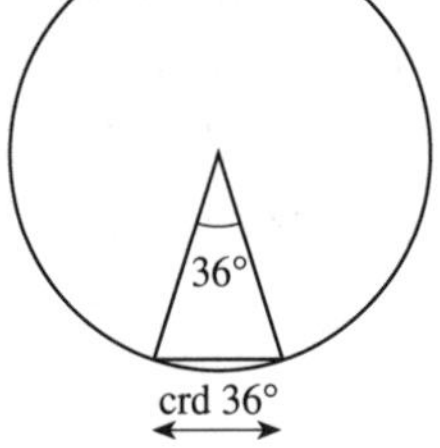

Figure 5.2

Trigonometry

Arab mathematicians inherited their knowledge of trigonometry from a variety of sources. In 150 BC in Alexandria, Hipparchus is believed to have produced the first trigonometric tables in order to study astronomy. These were tables of chords of a circle of unit radius. Figure 5.2 shows one of these chords.

Then, in AD 100, Ptolemy used this table to construct his own table of chords.

Activity 5.2   *The trigonometry of chords*

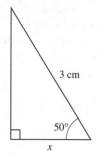

Figure 5.3a

1 Ptolemy's chord table contained entries like crd $36° = 0;37,4,55$ where the numbers are in sexagesimal notation.

a Use modern trigonometric methods to find the true value of crd $36°$, and determine whether Ptolemy's value is accurate, correct to three sexagesimal places.
b Ptolemy's tables could be regarded as three-figure sexagesimal tables. What would the fourth figure have been?

2 How would you use tables of chords to calculate the length x in the right-angled triangle in Figure 5.3a?

3 Explain how you could use chord tables to find x in Figure 5.3b. (Don't calculate x. It is a strategy which is required.)

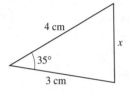

Figure 5.3b

You can see, from question **3** of Activity 5.2, that the trigonometry of chords is inconvenient when you come to deal with triangles which are not right-angled.

From AD 400 to AD 700, the Indian mathematicians continued the development of trigonometry, getting close to modern trigonometry. They used figures similar to Figure 5.4, in which the line AMB was called 'samasta-jya', the bow string, later abbreviated to 'jya'. By calculating AM, OM and MC, they developed tables for $\sin \alpha$, $\cos \alpha$ and $1 - \cos \alpha$, called vers α.

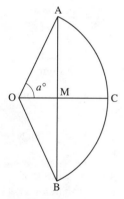

Figure 5.4

There is a curious story about the derivation of the term 'sine'. The word 'jya', when translated into Arabic, became the meaningless word 'jyb' – written, in the Arab custom, with consonants only. When this term was translated into Latin it was read as 'jaib', meaning bosom! The word 'sinus', meaning a fold in a toga, was used instead to refer to the sine.

The Arab mathematicians were able to develop tables of increasing accuracy as the following activity shows.

Activity 5.3   *Compiling trigonometric tables*

In this activity, the only operations for which you should use your calculator are the usual four arithmetic operations, plus taking a square root, because they were the only facilities available at this time. These operations were performed by hand and were time consuming; you have the advantage of using a calculator with a memory.

1 The Arabs calculated the sines and cosines of $30°$, $45°$, $60°$ using expressions familiar to you in terms of square roots. Write down these values.

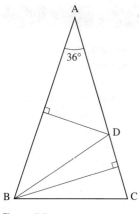

Figure 5.5

2   The value of $\cos 36°$ was found from Figure 5.5, in which the triangles ABC and BCD are isosceles.

a   By letting $BC = 1$ unit, and writing down expressions for AB and AC, show that $2\cos 36° = 2\cos 72° + 1$.

b   Use the relation $\cos 2\alpha = 2\cos^2 \alpha - 1$ and the result in part **a** to show that $\cos 36° = \frac{1}{4}\left(1 + \sqrt{5}\right)$.

c   Show further that $\cos 18° = \sqrt{\frac{1}{8}\left(5 + \sqrt{5}\right)}$.

d   Show how to find the value of $\sin 18°$.

3   The Arabs knew the formulae for $\sin(\alpha + \beta)$ and $\cos(\alpha + \beta)$ so they could calculate the value of $\sin 12°$. How?

They then used the halving technique, shown in questions **2c** and **2d**, to find $\sin 6°$, $\sin 3°$, $\sin 1\frac{1}{2}°$ and $\sin \frac{3}{4}°$ and then approximated by interpolation to find $\sin 1°$.

4   Given the values of $\sin 1\frac{1}{2}°$ and $\sin \frac{3}{4}°$ from your calculator, use linear interpolation to approximate to $\sin 1°$. How accurate is your answer?

A way of improving on the result for $\sin 1°$ would be to use a quadratic or a cubic approximation for the graph of $y = \sin x°$ between $x = \frac{3}{4}$ and $x = 1\frac{1}{2}$, but a better method is the one used by al-Kashi in the early 15th century.

Al-Kashi used the formula $\sin 3\alpha = 3\sin \alpha - 4\sin^3 \alpha$ together with the known value of $\sin 3°$ to calculate $\sin 1°$. Let $x = \sin 1°$. Then $3x - 4x^3 = \sin 3°$. He then re-wrote this equation in the form $x = \frac{1}{3}\left(\sin 3° + 4x^3\right)$ and used the $x_{n+1} = f(x_n)$ method which you met in the *Iteration* unit in Book 4 to find x.

5   a   Use this method, together with the value of $\sin 3°$ from your calculator, to find $\sin 1°$. Take $\frac{1}{3}\sin 3°$ as your first approximation. How many iterations do you need before the value of $\sin 1°$ does not change in its ninth place?

b   Similarly, start from $\sin 36°$ to find $\sin 12°$. Why is this method less efficient?

c   What happens if you try to find $\sin 30°$ from $\sin 90°$ using this method?

You can see that, with the arithmetic operations that the Arabs could do:
- compiling a set of trigonometric tables was an enormous undertaking
- choosing an efficient method for calculating a specific result was important.

However, once again it is worth thinking how much easier it was to construct tables with their 'modern' number system than with the earlier Babylonian system.

Arab recreational mathematics

Problems involving indeterminate equations and arithmetic and geometric series – both of which appear in various Indian texts whose origins may go back to the beginning of the Christian era – appear in early Arab mathematics, and form part of the subject of recreational mathematics of the Arabs. The two most popular examples of the time were the 'chess-board' and the 'hundred fowls' problems. For the chess-board problem, al-Khwarizmi, whose name gave mathematics the word 'algorithm', wrote a whole book, and the second problem formed part of the introduction to Abu Kamil's treatment of indeterminate equations.

The Arabs

This problem is believed by one Arab author to have originated, like chess, in India.

1   The man who invented chess was asked by his grateful ruler to demand anything, including half the kingdom, as reward. He requested that he be given the amount of grain which would correspond to the number of grains on a chess board, arranged in such a way that there would be one grain in the first square, two in the second, four in the third, and so on up to the 64 squares. The ruler first felt insulted by what he thought was a paltry request, but on due reflection realised that there was not sufficient grain in his whole kingdom to meet the request. The total number of grains required was 18 446 744 073 709 551 615. How did the chess-inventor find this number?

2   a   An example of an indeterminate equation in two unknowns is $3x + 4y = 50$, which has a finite number of positive whole-number solutions for (x, y). For example, $(14, 2)$ satisfies the equation, as does $(10, 5)$. Find all the other pairs.
b   Why do you think such an equation is called 'indeterminate'?

3   The 'hundred fowls' problem, which involves indeterminate equations, is found in a number of different cultures. The original source may have been India or China but it was probably transmitted to the West via the Arabs. Here is an Indian version which appeared around AD 850.

> Pigeons are sold at the rate of 5 for 3 rupas, cranes at the rate of 7 for 5 rupas, swans at the rate of 9 for 7 rupas and peacocks at the rate of 3 for 9 rupas. How would you acquire exactly 100 birds for exactly wrupas?

The author gives four different sets of solutions. Write an algorithm to find the solutions, turn it into a program and run it. How many solutions are there?

Before you run your program, refer to the hints section.

4   Interest in indeterminate equations in both China and India arose in the field of astronomy where there was a need to determine the orbits of planets. The climax of the Indian work in this area was the solution of the equations

$$ax^2 \pm c = y^2 \text{ and in particular } ax^2 + 1 = y^2$$

where solutions are sought for (x, y).

A version of this problem is first attributed by the Arab mathematicians to Archimedes who posed a famous problem known as the cattle problem, which reduced to finding a solution y of $x^2 = 1 + 4\,729\,494 y^2$ which is divisible by 9314. You can read about the cattle problem in, for example, *The history of mathematics: a reader*, by Fauvel and Gray, page 214, but don't try to solve it. These equations are now known as Pell equations, named by Euler by mistake after an Englishman, John Pell (1610–1685), who didn't solve the equation! The study of Pell equations has interested mathematicians for centuries. In 1889, A H Bell, a civil engineer from Illinois and two friends computed the first 32 digits and the last 12 digits of the solution, which has 206 531 digits.

Brahmagupta (born in AD 598) solved the Pell equation $8x^2 + 1 = y^2$ using the following method, called the cyclic method. It is presented, without explanation, rather like a recipe.

Brahmagupta first found a solution $x = 1$ and $y = 3$. He then carried out the process represented in Table 5.1.

Smaller root	Larger root	
1	3 ⎱	Multiply cross-wise and add to find a new value of x, that is, $x = 6$. The value of y is then found, that is, $y = 17$.
1	3 ⎰	
1	3 ⎱	Repeating the process gives $x = 35$, $y = 99$.
6	17 ⎰	
1	3 ⎱	And so on.
35	99 ⎰	

Table 5.1

a Use the algorithm to generate two further solutions to $8x^2 + 1 = y^2$.

b Investigate whether this method of solution applies to other Pell equations. For instance, for the equation $2x^2 + 1 = y^2$, a starting point is $x = 2$, $y = 3$.

The Arab mathematicians exploited their new number system for other numerical problems of a more recreational type. There is almost a parallel with present-day computing, when it is now possible, using computers, to tackle problems which were impossible before.

Activity 5.5   Amicable numbers

1 A pair of natural numbers, M and N, are defined as amicable if each is equal to the sum of the proper divisors of the other: that is, all the divisors of the other number, including 1, but not itself. The smallest pair of amicable numbers is 220 and 284. Check that this pair of numbers is amicable.

2 Thabit ibn Qurra provided the following algorithm for deriving pairs of amicable numbers. Using modern notation, let p, q, and r be distinct primes given by $p = 3 \times 2^{n-1} - 1$, $q = 3 \times 2^n - 1$, $r = 9 \times 2^{2n-1} - 1$, where n is an integer greater than 1. Then $M = 2^n pq$ and $N = 2^n r$ is a pair of amicable numbers.

a Find the first three pairs of amicable numbers given by this rule.

b Prove that numbers M and N of this form are amicable.

Thabit obtained only the first pair of amicable numbers from his rule. It was a later Arab mathematician, Ibn al-Banna, who found the next pair by applying the rule. Some 600 years after Thabit, the French mathematician Fermat (1601–1655) rediscovered the rule to find the next pair. Unfortunately, Thabit's rule has only limited application. It misses out the second smallest pair of amicable numbers, 1184 and 1210, which was overlooked by all the famous 'amicable number chasers' – to be discovered in 1866 by an Italian schoolboy! Fortunately, there are other approaches. The German mathematician Euler (1707–1783) found more than 60 pairs, using methods he developed himself – methods that still form the basis of present-day search techniques.

Algorithms for finding solutions to equations

The following two activities illustrate two ideas which originated in China and which were used by Arab mathematicians and handed on to the West. The first is an example of a numerical solution of higher-order equations by a Chinese method, which culminated in the work of Chin Chiu Shao in 1247 entitled *Chiu-chang suan-shu*, which appeared in its simpler versions in Arab mathematics about the same time. Nowadays this method is known as 'Horner's method'.

In 1819, an English schoolteacher and mathematician, W G Horner, published a numerical method of finding approximate values of the roots of equations of the type $f(x) = a_0 x^n + a_1 x^{n-1} + \ldots + a_{n-1} x + a_n = 0$. The procedure that Horner re-discovered is identical to the computational scheme used by the Chinese over 500 years earlier.

Here is an example in which this method is used to solve $x^2 + 252x - 5292 = 0$.

First, find that there is a root between 19 and 20. Then make the substitution $y = x - 19$ to obtain $(y+19)^2 + 252(y+19) - 5292 = 0$, or $y^2 + 290y - 143 = 0$, which you know has a solution between 0 and 1. Now say that

$$y \approx \frac{143}{(1+290)} = 0.491, \text{ so that } x \approx 19.491.$$

To find the cube root of 12 978 (that is, to solve the equation $x^3 - 12\,978 = 0$), first note that there is a solution between 20 and 30, so you substitute $y = x - 20$ to obtain the equation $y^3 + 60y^2 + 1200y - 4978 = 0$, which has a solution between 0 and 10. Now check that this equation has a solution between 3 and 4, so write $z = y - 3$ and obtain the equation $z^3 + 69z^2 + 1587z - 811 = 0$. Then you can say that $z \approx \dfrac{811}{1+69+1587} = 0.49$, so that $x \approx 23.49$.

Activity 5.6   Horner's method

Work in a group for this activity. Be prepared to make a short oral report on question 3 at your next review.

1 Use Horner's method to find the square root of 71 824.

2 Use Horner's method to solve the equations
a $x^2 - 23x - 560 = 0$
b $x^6 + 6000x = 191\,246\,976$.

3   Write notes on the accuracy of this method. For example, when is the approximation good, and when is it bad?

Activity 5.7   *The rule of 'double false'*

In the Chinese classic, *Chui-chang suan-shu* (Nine chapters on the mathematical arts), from the beginning of the Christian era, is the following problem.

> A tub of full capacity 10 tou contains a certain quantity of coarse, that is, husked, rice. Grains, that is, unhusked rice, are added to fill the tub. When the grains are husked it is found that the tub contains 7 tou of coarse rice altogether. Find the original amount of rice in the tub.

Assume that 1 tou of grains yields 6 sheng of coarse rice, where 1 tou is equal to 10 sheng and 1 sheng is approximately 0.2 litres.

The suggested solution is as follows.

> If the original amount of rice in the tub is 2 tou, a shortage of 2 sheng occurs; if the original amount of rice is 3 tou, there is an excess of 2 sheng. Cross-multiply 2 tou by the surplus 2 sheng, and then cross-multiply 3 tou by the deficiency of 2 sheng, and add the two products to give 10 tou. Divide this sum, that is 10, by the sum of the surplus and deficiency to get the answer: 2 tou and 5 sheng.

1   Take x_1 and x_2 to be the preliminary, incorrect, guesses of the answer, x. Take e_1 and e_2 to be the errors: that is, the shortage and surplus, arising from these guesses. Express in symbolic form the algorithm used in the solution quoted.

Questions **2** and **3** interpret this solution geometrically, in linear and other forms.

2   The linear case: In Figure 5.6a, x_1 and x_2 are incorrect guesses for x, and e_1 and e_2 are the consequent errors. Derive the expression obtained for x in question **1**.

3   The non-linear case: In Figure 5.6b, let x_1 and x_2 be numbers which lie close to and on each side of a root x of the equation $f(x) = 0$. You can find an approximate solution for the root x in two stages.
a   Apply the formula that you obtained in question **1** to find x_3.
b   Repeat the process, choosing the appropriate pair, x_1, x_3 or x_3, x_2 so that one is on either side of the solution of the equation. Use this method to find the approximate root which lies in the range between 1 and 2, correct to two places of decimals, of the equation $x^3 - x - 1 = 0$.

This method was called the **rule of double false**. In the linear case the solution for x is unique and independent of the choice of x_1 and x_2 if x_1 and x_2 are distinct. In the non-linear case, the solution is approximate. If you apply the method iteratively, as in the example above, a closer and closer solution may be found. The number of iterations needed depends on the accuracy of your initial approximations!

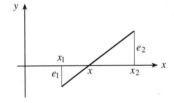

Figure 5.6a

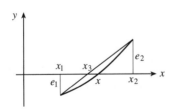

Figure 5.6b

The rule of double false provides one of the oldest methods of approximating to the real roots of an equation. Variations of it are found in Babylonian mathematics, in Greek mathematics in Heron's method for the extraction of a square root, and in an

Indian text. For over a thousand years, the Chinese used this method, refining and extending it, and then passed it on to the Arabs who in turn gave it to the Europeans. And the method is still used in modern numerical methods. Its Chinese antecedents, however, are not well known in the West.

Arab algebra

Perhaps the best-known mathematician of the Arab era was al-Khwarizmi. His work, *Arithmetic*, introduced the Hindu-Arabic system of numerals to the Arabs, and thence to western Europe. Unfortunately, only translations survive.

The word *al-jabr* appears frequently in Arab mathematical texts that followed al-Khwarizmi's influential *Hisab al-jabr w'al-muqabala*, written in the first half of the 9th century. Two meanings were associated with *al-jabr*. The more common one was 'restoration', as applied to the operation of adding equal terms to both sides of an equation, so as to remove negative quantities, or to 'restore' a quantity which is subtracted from one side by adding it to the other. Thus an operation on the equation $2x + 5 = 8 - 3x$ which leads to $5x + 5 = 8$ would be an illustration of al-jabr.

The less common meaning was multiplying both sides of an equation by a certain number to eliminate fractions. Thus, if both sides of the equation $\frac{9}{4}x + \frac{1}{8} = 3 + \frac{15}{8}x$ were multiplied by 8 to give the new equation $18x + 1 = 24 + 15x$, this too would be an instance of al-jabr. The common meaning of 'al-muqabala' is the 'reduction' of positive quantities in an equation by subtracting equal quantities from both sides. So, for the second equation above, applying al-muqabala would give successively

$$18x + 1 = 24 + 15x$$
$$18x - 15x + 1 - 1 = 24 - 1 + 15x - 15x$$

and $3x = 23$ or $x = \frac{23}{3}$.

Eventually the word al-jabr came to be used for algebra itself.

Notation

The Arabic word 'jadhir' meaning 'root' was introduced by al-Khwarizmi to denote the unknown x in an equation. With the use of terms such as 'number' for constant, 'squares' for x^2, 'kab' (cube) for x^3, or combinations of these terms, the Arabs were able to represent equations of different degrees. For example, you can translate 'one square cube and three square square and two squares and ten roots of the same equal twenty' into modern notation as $x^5 + 3x^4 + 2x^2 + 10x = 20$.

Al-Khwarizmi distinguished six different types of equations which he then proceeded to solve, providing both numerical and geometrical solutions.

The six different types of equation were
- squares equal to roots $ax^2 = bx$
- squares equal to numbers $ax^2 = c$
- roots equal to numbers $bx = c$
- squares and roots equal to numbers $ax^2 + bx = c$
- squares and numbers equal to roots $ax^2 + c = bx$
- roots and numbers equal to squares $bx + c = ax^2$

where a, b, and c are positive integers in each case.

The solutions of one equation, 'one square and ten roots of the same equal thirty nine (dirhems)' or $x^2 + 10x = 39$, is interesting historically since it recurs in later Arab and medieval European texts. Al-Khwarizmi's numerical and one of the geometric solutions 'completing the square' are quoted in full. They now constitute a standard repertoire in algebra.

Here is al-Khwarizmi's solution for the equation $x^2 + 10x = 39$.

> You halve the number (i.e. the coefficient) of the roots, which in the present instance yields five. This you multiply by itself; the product is twenty-five. Add this to thirty-nine; the sum is sixty four. Now take the root of this, which is eight, and subtract from it half the number of the roots, which is five; the remainder is three. This is the root of the square which you sought for; the square itself is nine.

Refer to Figure 5.7 for al-Khwarizmi's geometric solution.

> We proceed from the quadrate A B, which represents the square. It is our business to add to it the ten roots of the same. We halve for this purpose the ten, so that it becomes five, and construct two quadrangles (rectangles) on two sides of the quadrate A B, namely, G and D, the length of each of them being fives, ... whilst the breadth of each is equal to a side of the quadrate A B. Then a quadrate remains opposite the corner of the quadrate A B. This is equal to five multiplied by five; this five being half of the number of the roots which we have added to each of the two sides of the first quadrate. Thus we know that the first quadrate, which is the square, and the two quadrangles on its sides, which are the ten roots, make together thirty-nine. In order to complete the great quadrates, there wants only a square of five multiplied by five, or twenty-five. This we add to thirty-nine, in order to complete the great square S H. The sum is sixty-four. We extract its root, eight, which is one of the sides of the great quadrangle. By subtracting from this the same quantity which we have before added, namely five, we obtain three as the remainder. This is the side of the quadrangle A B, which represents the square; it is the root of this square, and the square itself is nine.

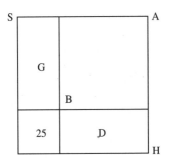

Figure 5.7

1 Explain in modern notation both solutions for the equation $x^2 + 10x = 39$ offered by al-Khwarizmi. Notice that al-Khwarizmi ignores the negative solution.

The geometric solution became more abstract and rigorous as Arab mathematicians mastered the works of Euclid and Apollonius. This is evident if you compare al-Khwarizmi's work with that of a later Arab mathematician, Abu Kamil.

2 Study the two mathematicians' geometric solutions, translated into modern notation, to the problem

> 'A square and twenty one (dirhems) equal ten roots.'

That is, solve $x^2 + 21 = 10x$.

a Here is a description of al-Khwarizmi's solution (see Figure 5.8).

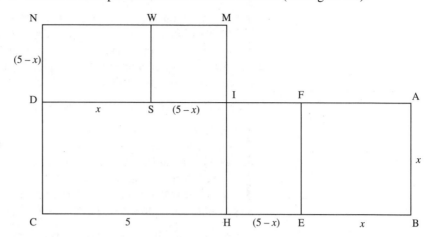

Figure 5.8

Beginning with the square ABEF of side x units, extend BE to H and then C and AF to I and then D such that ID = HC = 5 units and FI = EH = $5 - x$ units.

Notice that it follows that H is the mid-point of BC.

Therefore, the area of CDFE = $x(5 - x) + 5x = 10x - x^2 = 21$.

Next complete the square HMNC, which has an area of 25 units, by extending HI to M and CD to N such that DN = IM = EH = FI = $5 - x$.

Now drop a perpendicular from W to S such that the area of square ISWM = $(5 - x)^2$.

Since the area of SDNW = the area of EHIF = $x(5 - x)$, it follows that the area of square HMNC = the area of CDFE + the area of ISWM = 25.

Or the area of ISWM = 25 − 21 = 4 so that IS = FI = 2.

Subtracting FI from AI will give the root of the square ABEF, that is, $x = 3$. Adding FI to AI, that is, $5 + (5 - x)$, will give AS = 7.

Hence, $x = 3$ or 7.

b Here is Abu Kamil's solution (see Figures 5.9a and 5.9b).

Figure 5.9a is constructed in the following way. Beginning with the square ABDG of side x units, extend DB to L such that the area of ABLH is 21 units. By construction let BL be greater than DB. Also let DL be 10 so that the area of DLHG is $10x$.

Let P be the mid-point of DL. You need to show that $BL \times DB + (BP)^2 = (DP)^2$.

Figure 5.9b is constructed the following way. Beginning with the square DBAG of side x units, extend DB to W, with BW less than DB. Then let P be the mid-point of DW and draw the square PWLN.

Abu Kamil does not show this result but quotes the result from Euclid's *Elements*, *Book II*, proposition 5.

> If a straight line be cut into equal and unequal segments, the rectangle contained by the unequal segments of the whole together with the square on the straight line between the points of section is equal to the square on the half.

The proof given below uses a demonstration of Heron of Alexandria which Abu Kamil must have known.

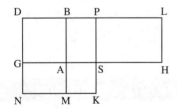

Figure 5.9a

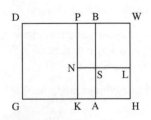

Figure 5.9b

An Arabic commentary by al-Naziri on Heron's work quotes Euclid's *Elements, Book II*, propositions 1, 3 and 2 in support of the results shown as (1), (3) and (5), respectively.

Given: $BL = BP + PL$ where P is the mid-point of DL.

To show that: $BL \times DB + (BP)^2 = (DP)^2$.

Proof: $DB \times BL = DB \times (BP + PL)$.   (1)

Add $(BP)^2$ to both sides of this equation and substitute DP for PL to get
$DP \times DB + BP \times DB + (BP)^2 = BL \times DB + (BP)^2$.   (2)

But $BP \times DB + (BP)^2 = (DB + BP) \times BP = DP \times BP$.   (3)

Combining the last two equations gives
$DP \times BP + DP \times DB - BL \times DB + (BP)^2$.   (4)

Also $DP \times BP + DP \times DB = DP \times (BP + DB) = (DP)^2$.   (5)

Combining the last two equations gives the final result
$BL \times DB + (BP)^2 = (DP)^2$.

To continue with Abu Kamil's solution, in Figure 5.9a, $BL \times DB + (BP)^2 = (DP)^2$ where $(DP)^2 = 5^2 = 25$, and $BL \times DB =$ area of ABLH $= 21$.
So $(BP)^2 = 4$ or $BP = 2$.
Therefore $DP - BP = DB = x = 5 - 2 = 3$ or the area of square ABDG is 9.
To find the other root of x, which is the side of square ABDG in Figure 5.9b, a similar argument, using another result from Euclid, gives
$WB \times BD + (BP)^2 = (PD)^2$ where $(PD)^2 = 25$, $WB \times BD = 21$ and $(BP)^2 = 4$.
So $PD + BP = BD = x = 5 + 2 = 7$ or the area of square ABDG is 49.

3 Express Euclid's rule algebraically and explain why it works.

4 Describe the two solutions in your own words.

5 What differences do you notice between the two approaches?

Solution of cubic equations

In his book *Algebra*, Omar Khayyam explored the possibility of using parts of intersecting conics to solve cubic equations. Traces of such approaches are found in the works of earlier writers such as Menaechmus (circa 350 BC) and Archimedes (287–212 BC), and Omar Khayyam's near contemporary al-Hayatham (circa 965–1039). They observed, for quantities a, b, c and d, that if $\dfrac{b}{c} = \dfrac{c}{d} = \dfrac{d}{a}$, then
$\left(\dfrac{b}{c}\right)^2 = \left(\dfrac{c}{d}\right)\left(\dfrac{d}{a}\right) = \dfrac{c}{a}$ or $c^3 = b^2 a$. If $b = 1$, you can calculate the cube root of a if c and d exist so that $c^2 = d$ and $d^2 = ac$.

Omar Khayyam re-discovered and extended the geometric argument implicit in this algebra. If you think of c and d as variables and of a as a constant, then the equations $c^2 = d$ and $d^2 = ac$ are the equations of two parabolas with perpendicular axes and the same vertex. This is illustrated in Figure 5.10. The two parabolas have the same vertex B, with axes $AB = a$ and $CB = b = 1$, and they intersect at E. In the rectangle BDEF, $BF = DE = c$ and $BD = FE = d$. Since AB is a line segment and the point E lies on the parabola with vertex B and axis AB, the rectangle BDEF has the property that $(FE)^2 = AB \times BF$ or $d^2 = ac$.

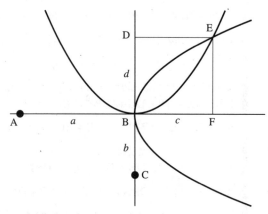

Figure 5.10

Similarly, for the other parabola with vertex B and axis BC, $(BF)^2 = CB \times BD$ or $c^2 = bd = d$. Combining the equations of these two parabolas gives $c^3 = a$. Therefore, DE = BF is a root of $c^3 = a$.

Applying similar reasoning, Omar Khayyam extended his method to solve any third degree equation for positive roots. He solved 19 different types of cubic equation, all expressed with only positive coefficients. Five of these equations could easily be reduced to quadratic or linear equations. Each of the remaining 14 he solved by means of cubic equations.

Using modern notation, Omar Khayyam was in effect solving an equation such as $x^3 + x = 20$ by making the substitution $y = x^2$ to arrive at the pair of equations

$y = \dfrac{20 - x}{x}$ and $y = x^2$. He then solved these equations graphically.

Activity 5.9   Solving cubic equations

1   Use the geometric method of Omar Khayyam, to solve the equations
a $x^3 + x = 20$
b $x^3 = x^2 + 20$.

Omar Khayyam's achievement is typical of Arab mathematics in its application, in a systematic fashion, of geometry to solve algebraic problems. While he made no addition to the theory of conics, he did apply the principle of intersecting conic sections to solving algebraic problems. In doing so he not only exhibited his mastery of conic sections but also showed that he was aware of the practical applications of what was a difficult area of geometry to understand. What Omar Khayyam and those who came after him failed to do was to investigate algebraic solutions of cubic equations. Omar Khayyam apparently believed that there were no algebraic methods for solving general cubic equations.

One of Omar Khayyam's successors in terms of work on cubics was a mathematician called Sharaf al-Din al-Tusi, who was particularly interested in looking at the conditions on the coefficients of cubics which determine how many solutions the cubic has. He therefore classified the cubics differently, into those types with at least one positive solution and those types that may and may not have a positive solution depending on the exact coefficients.

The following activity illustrates his approach.

Activity 5.10   Graphical deductions

1 Al-Tusi notes that for an equation $x^3 + c = ax^2$ expressed in the form $x^2(a - x) = c$, whether the equation has a positive solution depends on whether the expression of the left-hand side reaches c or not. He then states that, for any value of x lying between 0 and a, $x^2(a - x) < \left(\frac{2}{3}a\right)^2\left(\frac{1}{3}a\right)$. Why is this true?

2 Show that a local maximum occurs for x in the expression $x^2(a - x)$ when $x_0 = \frac{2}{3}a$. There is no indication of how al-Tusi found this value. Suggest how he might have found it.

3 He then proceeds to make the following inferences:
- if $\frac{4}{27}a^3 < c$ no positive solution exists
- if $\frac{4}{27}a^3 = c$, there is only one positive solution, $x = \frac{2}{3}a$
- if $\frac{4}{27}a^3 > c$, two positive solutions x_1 and x_2 exist where $0 < x_1 < \frac{2}{3}a$ and $\frac{2}{3}a < x_2 < a$.

Explain why this is true.

Al-Tusi was also interested in finding numerical solutions to cubic equations – he was one of the Arab mathematicians who made use of Horner's method, which you met in Activity 5.7.

Conclusion

There has always been a recognition on the part of western historians of science and of ideas of the debt that the European Renaissance owed to ancient Greece. But the original contributions of the Arabs are often neglected or devalued, perhaps because they wrote in a language which was difficult for western scholars to read.

Two important conditions were satisfied by the Arab culture: they were 'the melting pot effect' and the willingness of the Arabs to take ideas from anywhere and develop them in an integrated way.

The Renaissance is the name given to the period from the 14th to 16th centuries in Europe when there was a revival of cultural thinking.

Reflecting on Chapter 5

What you should know

- some examples of inheritance problems
- the contributions made by al-Khwarizmi and Omar Khayyam
- that Arab mathematics is in part a synthesis of Greek mathematics and mathematics from the East
- how decimal numbers were represented in the Arab number system
- how sine tables were constructed
- Thabit's method for amicable numbers
- examples of how Arab mathematicians developed solutions of equations from geometry
- the derivation of the word 'algebra'
- al-Khwarizmi's classification of quadratic equations
- al-Khwarizmi's method of solving quadratics geometrically
- Omar Khayyam's method of solving cubic equations.

Preparing for your next review

- Reflect on the 'What you should know' list for this chapter. Be ready for a discussion on any of the points.
- Answer the following check questions.

1 Show how you would apply al-Khwarizmi's method to solving equations such as $2x^2 + 10x = 48$ or $\frac{1}{2}x^2 + 5x = 28$. See whether you can prove the method geometrically.

2 Summarise in not more than one page, the work of al-Khwarizmi and Omar Khayyam. Use the text in this unit, and any other sources available to you.

3 The Arab contribution to mathematics has often been perceived as one of guardian of Greek mathematics while the West was in the turmoil of the dark ages. Summarise your own understanding of the Arab contribution with this in mind.

4 Solve another version of the hundred fowls problem as given below.

A rooster is worth 5 coins, a hen 3 coins and 3 chicks 1 coin. With 100 coins we buy 100 fowls. How many roosters, hens and chicks are there?

Practice exercises for this chapter are on page 154.

Descartes

Introduction

You can think of the algebra of the 12th to 16th centuries in Europe as being based on Latin translations of Arabic work, and being broadly algorithmic in nature. It was not until the 16th century that the Greek mathematical texts became available in comprehensible Latin.

It was then that the transformation of algebra under the influence of Greek geometry begun by the Arabic mathematicians was continued in Europe, first by a French mathematician, Viète. Another Frenchman Descartes, by discarding some of Viète's ideas and building on others, was able to make the major breakthrough you will study in this unit.

In Chapter 6 of this unit you will learn how Descartes used the introduction of algebra to make Greek construction problems considerably easier to solve.

In Chapter 7 you will meet his methods, which show his full genius. Thanks to his mathematical ideas, an algebraic approach for all geometric problems became available, and the basis for modern algebraic geometry was laid.

In Chapter 8 Descartes's work is put into perspective.

René Descartes (1596–1650) was a Frenchman. He lived a few kilometres south of Tours, and also in Holland and Sweden. He has been called the father of modern philosophy.

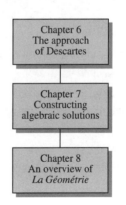

Chapter 6
The approach
of Descartes

Chapter 7
Constructing
algebraic solutions

Chapter 8
An overview of
La Géométrie

This unit is designed to take about 15 hours of your learning time. About half of this time will be outside the classroom.

Work through the chapters in sequence.

Many of the activities are in a different style from usual. They ask you to read a translation of Descartes's work, and to answer questions to ensure that you have understood it.

There are summaries and further practice exercises in Chapter 12.

Mathematical knowledge assumed

- the contents of Chapter 3 in *The Greeks* unit is particularly relevant to this unit
- for Activity 6.13, you will need to know some of the angle properties of circles; in particular you should know the 'angle at the centre is twice the angle at the circumference' and 'the angles in the same segments are equal' theorems.

6 The approach of Descartes

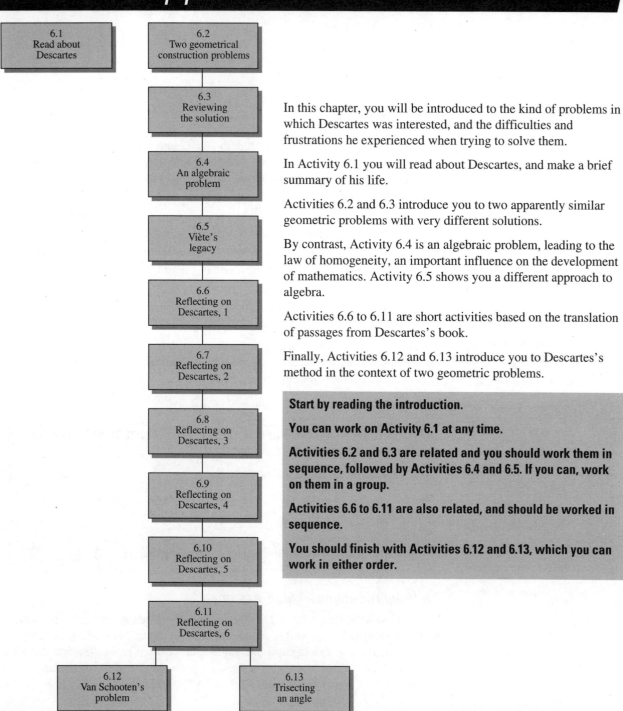

6.1
Read about
Descartes

6.2
Two geometrical
construction problems

6.3
Reviewing
the solution

6.4
An algebraic
problem

6.5
Viète's
legacy

6.6
Reflecting on
Descartes, 1

6.7
Reflecting on
Descartes, 2

6.8
Reflecting on
Descartes, 3

6.9
Reflecting on
Descartes, 4

6.10
Reflecting on
Descartes, 5

6.11
Reflecting on
Descartes, 6

6.12
Van Schooten's
problem

6.13
Trisecting
an angle

In this chapter, you will be introduced to the kind of problems in which Descartes was interested, and the difficulties and frustrations he experienced when trying to solve them.

In Activity 6.1 you will read about Descartes, and make a brief summary of his life.

Activities 6.2 and 6.3 introduce you to two apparently similar geometric problems with very different solutions.

By contrast, Activity 6.4 is an algebraic problem, leading to the law of homogeneity, an important influence on the development of mathematics. Activity 6.5 shows you a different approach to algebra.

Activities 6.6 to 6.11 are short activities based on the translation of passages from Descartes's book.

Finally, Activities 6.12 and 6.13 introduce you to Descartes's method in the context of two geometric problems.

Start by reading the introduction.

You can work on Activity 6.1 at any time.

Activities 6.2 and 6.3 are related and you should work them in sequence, followed by Activities 6.4 and 6.5. If you can, work on them in a group.

Activities 6.6 to 6.11 are also related, and should be worked in sequence.

You should finish with Activities 6.12 and 6.13, which you can work in either order.

Introduction

LA
GEOMETRIE.
LIVRE PREMIER.

*Des problefmes qu'on peut conftruire fans
y employer que des cercles & des
lignes droites.*

Ou s les Problefmes de Geometrie fe
peuuent facilement reduire a tels termes,
qu'il n'eft befoin par aprés que de connoi-
ftre la longeur de quelques lignes droites,
pour les conftruire.

Et comme toute l'Arithmetique n'eft compofée, que
de quatre ou cinq operations, qui font l'Addition, la
Souftraction, la Multiplication, la Diuifion, & l'Extraction des racines, qu'on peut prendre pour vne efpece
de Diuifion : Ainfi n'at'on autre chofe a faire en Geometrie touchant les lignes qu'on cherche, pour les pre-
parer a eftre connuës, que leur en adioufter d'autres, ou
en ofter, Oubien en ayant vne, que ıe nommeray l'vnité.
pour la rapporter d'autant mieux aux nombres, & qui
peut ordinairement eftre prife a difcretion, puıs en ayant
encore deux autres, en trouuer vne quatriefme, qui foit
à l'vne de ces deux, comme l'autre eft a l'vnité, ce qui eft
le mefme que la Multiplication; oubien en trouuer vne
quatriefme, qui foit a l'vne de ces deux, comme l'vnité

Commĕt
le calcul
d'Ari-
thmeti-
que fe
rapporte
aux ope-
rations de
Geome-
trie.

P p   eft

This is the title page of one of the most important mathematical works of all time.
Strangely, it was not intended as a mathematical work in its own right, since the
French philosopher René Descartes wrote it as an appendix to illustrate the method
of reasoning which he developed in his famous philosophical key-work *Discours de
la méthode pour bien conduire sa raison et chercher la vérité dans les sciences*
(Discourse on the method to guide one's reasoning and to find the truth in the
sciences). The *Discours* was published in 1637 in Leiden, a small university town in
Holland. Descartes moved in 1628 to Holland because of the greater freedom of
thought that prevailed there as opposed to the situation in France, where the
monarchy, nobility and clergy created a climate which restricted original thinking.

Descartes

1   Read a book, such as *A concise history of mathematics* by D J Struik, and summarise in about 10 lines the most important details about Descartes's life.

The appendix of Descartes's book, *La Géométrie*, became a major influence on mathematics after the Leiden professor of mathematics, Frans van Schooten, published a Latin translation with extensive commentaries and explanations.

But why was this work so important? What revolution came about when mathematicians began to understand and elaborate on Descartes's ideas?

The main ideas, which now seem simple, are these.

- Descartes used algebra to describe geometrical objects and to solve geometrical problems.
- He also used geometry and graphical methods to solve algebraic problems.

Describing geometrical objects and ideas in the language of algebra is now commonplace. Every student thinks of a^3 as the volume of a cube with edge length a, and can verify that you can make a larger cube, of edge $2a$, with eight of these smaller cubes. It feels quite normal to prove this geometric statement by using algebra. For example, eight smaller cubes have volume $8a^3$; one cube of edge $2a$ has volume $(2a)^3$; and the proof amounts to observing that $(2a)^3 = 8a^3$.

This algebraic thinking was a major step forward. It is difficult for us to understand the intellectual boldness of this step in 1637, when the usual way of thinking about geometry was that of the ancient Greeks. To recognise the ingenuity of Descartes's work you must remind yourself about the mathematics before Descartes.

The difficulty of geometric constructions

In Chapter 3 in *The Greeks* unit you learned about the Greeks' classical geometrical construction problems and the strict rules that their solutions had to obey. In the next two activities you will re-visit two problems to re-create some of these difficulties for yourself. You will then begin to understand what it was that frustrated Descartes to such an extent that he created a new way of solving these problems.

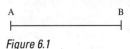

Figure 6.1

If you can, work in a group, especially for question 4.

Figure 6.1 shows two points, A and B, in a plane.

1   Use a straight edge and a pair of compasses to construct the set of points such that the distance of each point X of the set from A is equal to its distance from B.

2   Show that each point X of your set satisfies XA = XB. Show also that no other point can be a member of your set. If you can, prove that your solution is correct.

A proof that your solution is correct is only possible if you were able to draw the set of points. But how did you get the idea that you must have this particular set of points: how did you arrive at your solution?

If you have not been able to answer question **2**, do not worry. You will return to it later. But first work on question **3**. It appears to resemble question **1**, but it has a quite different solution. Again it is based on the two points A and B in Figure 6.1.

3 Construct the set of points such that the distance of each point X from A equals twice its distance from B. If you cannot find all the points that satisfy the condition, trying finding some individual points that do.

4 If you found or guessed a solution to question **3**, prove that your set of points satisfies the condition.

It is quite likely that you cannot find the solution set for question **3**. Although questions **1** and **3** are very similar, the solutions and the strategies to find them are quite different, unless you introduce a coordinate system and use algebra. There is no single straight-forward geometric method for solving both problems.

A construction problem

Here is a typical solution to a geometrical construction problem to illustrate the point made in the last paragraph. Return to question 3 of Activity 6.2.

The problem

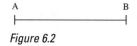

Figure 6.2

A and B are two given points. Construct the set of points X such that the distance of X from A is twice the distance of X from B, that is, $AX = 2BX$.

Experimenting

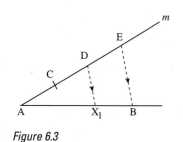

Figure 6.3

First, try to find, somehow, some individual points X that satisfy the condition.

The point X_1, shown in Figure 6.3, is on AB, two-thirds of the way from A, so it satisfies the condition. To construct X_1 take a half-line m, which passes through A at an angle to AB. On m, starting from A, mark three equal line-segments AC, CD and DE. Draw EB. Draw a line through D parallel to EB; this line intersects AB in the required point X_1. Notice that AX_1 is two-thirds of AB because triangles ABE and AX_1D are similar.

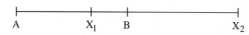

Figure 6.4

Figure 6.4 shows that on AB produced there is a second point X_2 which satisfies the condition $AX = 2BX$, since, if BX_2 equals AB, then $AX_2 = 2BX_2$.

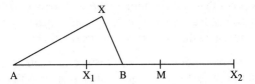

Figure 6.5

Figure 6.5 shows that there are other points, not on AB, that lie on a curve that looks as though it may be a circle. After drawing a few of these points you may be ready to make a hypothesis.

Hypothesis

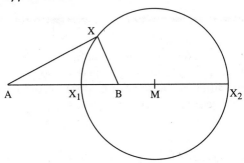

Figure 6.6

The set of points X that satisfy the condition that $AX = 2BX$ is a circle, as shown in Figure 6.6. The centre M of the circle is on AB produced, such that $MB = \frac{1}{3}AB$, and the radius is $\frac{2}{3}AB$.

Now that you have investigated the problem and made a hypothesis, try to prove that your hypothesis is correct.

Proof

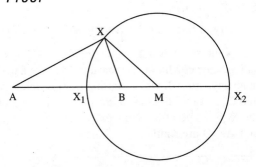

Figure 6.7

For each point X on the circle shown in Figure 6.7
- $\angle AMX = \angle XMB$
- $AM : XM = 2 : 1$ (since $AM = AB + BM = \frac{4}{3}AB$ and $XM = $ radius $= \frac{2}{3}AB$)
- also $MX : MB = 2 : 1$ (why?)

So the triangles AMX and XMB are similar. Therefore $AX : XB = 2 : 1$.

1   Compare your own attempt at the solution of the $AX = 2BX$ problem, and the solution of some fellow students, with that presented in the text above. Which of the solutions do you judge to be the most apt and practical?

2   Why do you think it the most apt? In what respect is it more practical than the other solutions?

3   Did any of you use algebra in analysing the problem? Is algebra used in the solution presented in the text above?

4   In the light of the remarks made just after Activity 6.2 and at the beginning of this section, reflect briefly on the solution in the text to question **3** in Activity 6.2, comparing it with your solution to question **1** in Activity 6.2.

> Recall from Chapter 3, that a quadrature problem involved finding a square which was equal in area to another figure.

> Cardano and Tartaglia were both involved in the solution of cubic equations. If you would like to read about their rivalry, see 'Tartaglia, the stammerer' in the *Mathematics reader*.

Interest in the culture of the ancient Greeks was one of the leading themes in these times. Mathematicians had a keen interest in Greek geometry, and a Latin translation of Euclid's *Elements* was published in 1482. Geometrical construction problems, not least quadrature problems, were studied on a large scale.

You may have noticed that one of the important subjects in present day mathematics, algebra, has hardly been mentioned up to now in this unit. People used algebra in very early times, for example, in Mesopotamia in about 1800 BC, as you have seen in *The Babylonians* unit, in the later Greek period and in Arab mathematics. In 16th century Italy, through the work of Cardano, Tartaglia and others, algebra became valued and popular. Cardano, who wrote a famous algebra book, *Ars Magna*, in 1545, considered it to be a 'great art'. But, before Descartes, geometry and algebra were separate disciplines with almost nothing in common. The 17th century was to change that significantly.

The law of homogeneity

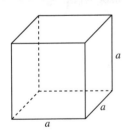

Figure 6.8

Consider a cube of side a, shown in Figure 6.8. Three numbers associated with this cube are the total length, L, of the edges of the cube, the total surface area, A, of the cube, the volume, V, of the cube.

1   For which value of a is $A = L$?

2   For which value of a is $V = L$?

3   For which value of a is $V + A + L = 90$?

Give your answers correct to two decimal places. Use a graphics calculator if necessary.

Descartes

Before Descartes, problems which involved comparing lengths, areas and volumes were inconceivable. It was not the difficulty of solving the equations which was the problem. People of that time could calculate the answers, if necessary to an accuracy of two decimal places, even without modern calculation aids.

In the years before Descartes, the French mathematician François Viète (1540–1603) was very influential. In his work on algebra, he set a geometric requirement for his variables. Freely interpreted, Viète's requirement amounted to the idea that lengths, areas and volumes cannot be added. Lengths can only be added to lengths, areas to areas and volumes to volumes. In short, you can only add, algebraically, quantities of the same dimensions, called homogeneous quantities. The same holds true for subtracting quantities from each other, but not for multiplying and dividing.

Viète's requirement is called **the law of homogeneity**.

Variables a and b are, from a geometrical standpoint, the lengths of line segments. The algebraic expression $a+b$ is equivalent to the line segment whose length is the sum of the lengths of the line segments a and b. Similarly, $a \times b$ is the area of the rectangle with sides a and b. The quantity a^3 represents a cube. However, $a^3 + 3b$ does not satisfy the law of homogeneity, because length is being added to volume.

Here is a translation of part of a chapter from Viète's writings.

> On the law of homogeneous quantities and the comparison of quantities in degrees and sorts. The first and general law with respect to equations or relations, that, since it is based on homogeneous matters, is called the law of the homogeneous quantities.
>
> Homogeneous things can only be compared to homogeneous things. This is because you have no idea how to compare heterogeneous things, as Adrastus said.
>
> Therefore, if a quantity is added to another quantity, the result is consequently homogeneous.
>
> If a quantity is subtracted from another quantity, the result is consequently homogeneous.
>
> If a quantity is multiplied with another quantity, the result is consequently heterogeneous with this and that quantity.
>
> If a quantity is divided by another quantity, the result is consequently heterogeneous.
>
> In antiquity they didn't consider this law and this caused the great darkness and blindness of the old analysts.

Viète's work in bringing algebra and geometry together was a great step forward, both in algebra and in geometry. Algebra gained a classical geometrical grounding. Mathematicians could now formulate geometrical problems algebraically.

However, Viète's law of homogeneity forced algebra into a geometrical straitjacket. Because of this law, expressions such as $a^3 + 6a^2 + 12a$ had no geometric meaning. Descartes's *La Géométrie* helped mathematicians to escape from this straitjacket.

As you saw in *The Arabs* unit, their work on algebra was carried out using special cases. For example, ABC could represent any triangle, but there was no equivalent way of representing all possible quadratic equations.

Viète introduced a notation in which vowels represented quantities assumed unknown, and consonants represented known quantities.

In modern notation he would have written $BA^2 + CA + D = 0$ for quadratic equations. However, he used medieval rather than the current notation for powers, such as '*A cubus, A quadratus*', and '*In*' (Latin) instead of '×', '*aequatur*' instead of 'equals'.

1 Translate the following equations from Viète's work into modern notation. Assume that the word 'plano' means 'plane'.
a *A* quadratus + *B*2 in *A*, aequatur *Z* plano
b *D*2 in *A* – *A* quadratus, aequatur *Z* plano
c *A* quadratus – *B* in *A*2, aequatur *Z* plano
d Explain how the three equations in parts **a**, **b** and **c** fit Viète's ideas about the law of homogeneity.
e Why would Viète have needed to solve each of the equations in parts **a**, **b** and **c** separately?

Descartes's breakthrough

When you cannot use algebra, you cannot use a generally applicable method to solve geometrical problems. If you can use algebra, a second difficulty emerges: how do you translate this algebraic solution into a geometric construction?

It was also impossible to solve the harder problems algebraically without violating the law of homogeneity.

It was his frustration with these difficulties that led Descartes to write *La Géométrie*, which was his response to a number of questions related to geometrical construction methods that had so far remained unanswered.
● Is it possible to find a generally applicable solution method for the problems that, until Descartes's time, had been solved in a rather unstructured manner?
● How can you tell whether or not it is possible to carry out a particular construction with a straight edge and a pair of compasses?
● What method can you use if it is not possible to carry out a construction with a straight edge and a pair of compasses?

Descartes worked to find a single strategy to solve these geometrical problems. The strategy that he gives in *La Géométrie* is an elaboration of his general philosophical thinking: a method leading to truth and certainty.

La Géométrie comprises three parts, which Descartes called 'books'.
● *Book I* 'Construction problems requiring only lines and circles'
● *Book II* 'About the nature of curves'
● *Book III* 'About constructions requiring conic sections and higher-order curves'

The next two sections consist of translations of a number of passages from *Book I.* In these translations the paragraphs are numbered 1 to 11.

Introduction of units

The first passage is the beginning of *Book I.*

> ### The geometry of René Descartes
>
> ### Book I
>
> #### Problems the construction of which requires only straight lines and circles
>
> 1   Any problem in geometry can easily be reduced to such terms that a knowledge of the lengths of certain straight lines is sufficient for its construction. Just as arithmetic consists of only four or five operations, namely, addition, subtraction, multiplication, division and the extraction of roots, which may be considered a kind of division, so in geometry, to find required lines it is merely necessary to add or subtract other lines;

Descartes comes straight to the point and introduces a close relationship between geometry and arithmetic.

Activity 6.6   Reflecting on Descartes, 1

1   According to Descartes, what are the basic variables that can be used to describe all geometric problems?

Descartes continues.

> 2   ... or else, taking one line which I shall call unity in order to relate it as closely as possible to numbers, and which can in general be chosen arbitrarily, and having given two other lines, to find a fourth line which shall be to one of the given lines as the other is to unity (which is the same as multiplication); or, again, to find a fourth line which is to one of the given lines as unity is to the other (which is equivalent to division); or, finally, to find one, two, or several mean proportionals between unity and some other line (which is the same as extracting the square root, cube root, etc., of the given line). And I shall not hesitate to introduce these arithmetical terms into geometry, for the sake of greater clearness.

Descartes immediately comes forward with a powerful idea: the introduction of a line segment, chosen arbitrarily, that he calls a unit. He uses this unit, for example, in the construction of mean proportionals.

In Activity 3.3 you were introduced to the problem of the mean proportional. Let a and b be two line segments. Determine the line segment x such that $a : x = x : b$.

Euclid solved this problem using a right-angled triangle. The hypotenuse of the right-angled triangle is $a + b$; the perpendicular to the hypotenuse, shown in

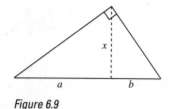

Figure 6.9

Figure 6.9, is constructed and produced to form a right-angled triangle. The mean proportional is the perpendicular from the right-angled vertex to the hypotenuse.

Activity 6.7 — *Reflecting on Descartes, 2*

1 What is the length of the mean proportional if you take *b* as the unit?

2 Express *x* and *y* in the following equations in terms of *a* and *b*.
a $1 : a = b : x$ **b** $a : 1 = b : x$
c $1 : x = x : a$ **d** $1 : x = x : y = y : a$

3 Where, in the Descartes's text, can you find the geometrical equivalents of the equations in question **2**?

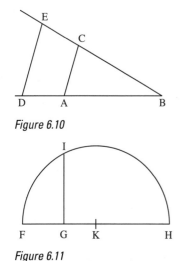

Figure 6.10

Here is the third extract from Descartes's text.

> **3** For example, let AB be taken as unity, and let it be required to multiply BD by BC. I have only to join the points A and C, and draw DE parallel to CA; then BE is the product of BD and BC.
>
> If it be required to divide BE by BD, I join E and D, and draw AC parallel to DE; then BC is the result of the division.
>
> If the square root of GH is desired, I add, along the same straight line, FG equal to unity; then, bisecting FH at K, I describe the circle FIH about K as a centre, and draw from G a perpendicular and extend it to I, and GI is the required root. I do not speak here of cube root, or other roots, since I shall speak more conveniently of them later.

Figure 6.11

Activity 6.8 — *Reflecting on Descartes, 3*

1 Prove that $BC = \dfrac{BE}{DB}$ (see Figure 6.10).

2 Prove that $GI = \sqrt{GH}$ (see Figure 6.11).

The law of homogeneity is relaxed

Continuing with the extract:

> **4** Often it is not necessary thus to draw the lines on paper, but it is sufficient to designate each by a single letter. Thus, to add the lines BD and GH, I call one *a* and the other *b*, and write $a + b$. Then $a - b$ will indicate that *b* is subtracted from *a*; *ab* that *a* is multiplied by *b*; $\dfrac{a}{b}$ that *a* is divided by *b*; *aa* or a^2 that *a* is multiplied by itself; a^3 that this result is multiplied by *a*, and so on, indefinitely. Again, if I wish to extract the square root of $a^2 + b^2$, I write $\sqrt{a^2 + b^2}$; if I wish to extract the cube root of

$a^3 - b^3 + ab^2$, I write $\sqrt[3]{a^3 - b^3 + ab^2}$, and similarly for other roots. Here it must be observed that by a^2, b^3, and similar expressions, I ordinarily mean only simple lines, which, however, I name squares, cubes, etc., so that I may make use of the terms employed in algebra.

Activity 6.9   Reflecting on Descartes, 4

1 Descartes introduces an efficient way of working so that it is no longer necessary to draw all line segments. What is this way of working?

2 From the above extract it is clear that Descartes introduces a kind of revolution. He does not obey the law of homogeneity. This is evident in one of the sentences in the text. Which sentence?

3 In the texts before this one he has, in some places, also pushed aside the law of homogeneity. Give two examples.

To continue:

5 It should also be noted that all parts of a single line should always be expressed by the same number of dimensions, provided unity is not determined by the conditions of the problem. Thus, a^3 contains as many dimensions as ab^2 or b^3, these being the component parts of the line which I have called $\sqrt[3]{a^3 - b^3 + ab^2}$. It is not, however, the same thing when unity is determined, because unity can always be understood, even where there are too many or too few dimensions; thus, if it be required to extract the cube root of $a^2b^2 - b$, we must consider the quantity a^2b^2 divided once by unity, and the quantity b multiplied twice by unity.

Activity 6.10   Reflecting on Descartes, 5

1 How does Descartes remedy the problem of having terms of different dimensions in one equation?

The solution strategy

At this point Descartes describes the strategy needed to solve a geometrical construction problem.

6 Finally, so that we may be sure to remember the names of these lines, a separate list should always be made as often as names are assigned or changed. For example, we may write, AB = 1, that is AB is equal to 1; GH = a, BD = b, and so on.

7 If, then, we wish to solve any problem, we first suppose the solution already effected, and give names to all the lines that seem needful for its construction, — to those that are unknown as well as to those that are known. Then, making no distinction between known and unknown lines, we must unravel the difficulty in any way that shows most naturally the relations between these lines, until we find it possible to express a single quantity in two ways. This will constitute an equation, since the terms of one of these two expressions are together equal to the terms of the other.

8   We must find as many such equations as there are supposed to be unknown lines; but if, after considering everything involved, so many cannot be found, it is evident that the question is not entirely determined. In such a case we may choose arbitrarily lines of known length for each unknown line to which there corresponds no equation.

9   If there are several equations, we must use each in order, either considering it alone or comparing it with the others, so as to obtain a value for each of the unknown lines; and so we must combine them until there remains a single unknown line which is equal to some known line, or whose square, cube, fourth power, fifth power, sixth power, etc., is equal to the sum or difference of two or more quantities, one of which is known, while the others consist of mean proportionals between unity and this square, or cube, or fourth power, etc., multiplied by other known lines. I may express this as follows:

$$z = b,$$
$$\text{or } z^2 = -az + b^2,$$
$$\text{or } z^3 = az^2 + b^2 z - c^3,$$
$$\text{or } z^4 = az^3 - c^3 z + d^4, \text{ etc.}$$

That is, z, which I take for the unknown quantity, is equal to b; or the square of z is equal to the square of b diminished by a multiplied by z; or, the cube of z is equal to a multiplied by the square of z, plus the square of b multiplied by z, diminished by the cube of c; and similarly for the others.

10   Thus, all the unknown quantities can be expressed in terms of a single quantity, whenever the problem can be constructed by means of circles and straight lines, or by conic sections, or even by some other curve of degree not greater than the third or fourth.

11   But I shall not stop to explain this in more detail, because I should deprive you of the pleasure of mastering it yourself, as well as of the advantage of training your mind by working over it, which is in my opinion the principal benefit to be derived from this science. Because, I find nothing here so difficult that it cannot be worked out by those who have a certain familiarity with ordinary geometry and with algebra, and who will consider carefully all that is set forth in this treatise.

This strategy consists of the following stages.
- Work as if you have already solved the problem. For example, sketch a possible solution.
- Assign letters to the known and the unknown line segments.
- Write the relations between the separate line segments as algebraic equations.

- By combining equations, eliminate all the unknown quantities except for one. (The problem does not have a finite number of solutions if the number of equations is less than the number of quantities.)
- 'Translate' the algebraic solution of the remaining unknown quantity into a geometrical construction.

Here is an example to illustrate this process.

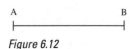

Figure 6.12

An example of Descartes's method

In Activity 6.2 you were asked to construct the set of points X such that the distances XA and XB are equal. The points A and B are shown in Figure 6.12.

A solution according to Descartes could be as follows:
- Sketch the situation with point X as a possible solution. Draw the perpendicular from X to AB. This line intersects AB in point S, as shown in Figure 6.13.
- The unknown quantities are AS and SX. Let AS = x and SX = y, and the known quantity AB = d.
- As AX = BX, you know that $\sqrt{x^2 + y^2} = \sqrt{(d-x)^2 + y^2}$.
- As you have only one equation, no elimination is necessary. It follows by squaring and simplifying that

$$x^2 + y^2 = d^2 - 2dx + x^2 + y^2$$
$$2dx = d^2$$
$$x = \tfrac{1}{2}d$$

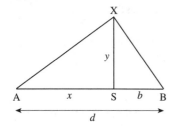

Figure 6.13

> The equation has two unknown quantities, indicating that there will not be one solution but a set of solutions. You will come back to this in the next chapter.

You can see that the solution in this case is the perpendicular on AB at a distance equal to $\tfrac{1}{2}d$ from A. However, such a 'translation' is not always that simple and can itself be a problem. You will study the translation from algebra to geometry in the next chapter.

Activity 6.11   Reflecting on Descartes, 6

1 Look back at the five points that summarise Descartes's strategy. Identify each of these stages in Descartes's solution.

2 Why did Descartes describe the part in paragraph 9 so extensively? To us this is clear immediately, isn't it?

3 In paragraph 9, Descartes uses the letters a to d for the known quantities and z for the unknown quantity. What system has he used for assigning these letters?

Activity 6.12   Van Schooten's problem

Descartes's *La Géométrie* was rather theoretical and lacked examples. When Frans van Schooten, from Leiden, translated it into Latin, he added clarifying comments and illustrative examples, including the following problem.

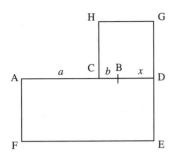

Figure 6.14

1 Figure 6.14 shows a straight line AB, with a point C on it. Produce AB to the point D such that the rectangle with sides AD and DB, is equal (in area) to the square with side CD.

Call the lengths AC, CB and BD, a, b and x.

1 Formulate an equation involving a, b and x.

2 Express x in terms of a and b.

Trisecting an angle

A classical problem from Greek mathematics was the problem of trisecting an angle.

Let O be an angle, shown in Figure 6.15. Construct two lines that divide the angle at O into three equal parts.

Descartes, who did not know that this problem cannot be solved with a pair of compasses and a straight edge, hoped to find a solution using his new method.

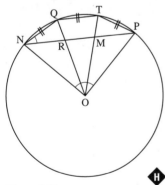

Figure 6.15

It was only in 1837 that a relatively unknown Frenchman, Pierre-Laurent Wantzel (1814–1848), proved that it was impossible to trisect an angle using only a straight edge and a pair of compasses.

The algebraic analysis of this problem is given in Activity 6.13. The geometric construction is left until Chapter 7.

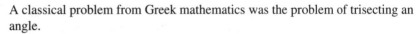

Activity 6.13   Trisecting an angle

Let O be the centre of a circle with radius 1. The points N, Q, T, P, R and M are as shown in Figure 6.16.

With the help of the figure, you can re-formulate the problem: 'Let NP be a chord. Find the length of chord NQ'.

1 By considering angles standing on the chord PQ, prove that $\angle NOQ = \angle QNR$.

2 Prove that triangle NOQ is similar to triangle QNR, and triangle ROM is similar to triangle QNR.

3 Let $NP = q$ and $NQ = z$. Prove that $RM = q - 2z$.

4 Use similar triangles to prove that $z^3 = 3z - q$.

Figure 6.16

Reflecting on Chapter 6

What you should know

● the major contributions made by Descartes and Viète to the development of mathematics

- the meaning of the law of homogeneity
- how to explain Descartes's approach to the construction of the bisector of the line AB.

Preparing for your next review

- Reflect on the 'What you should know' list for this chapter. Be ready for a discussion on any of the points.
- Answer the following check questions.

1 In *A history of mathematics (Second edition),* by Carl B Boyer and Uta C Merzbac, it states that

> Descartes substituted homogeneity in thought for homogeneity in form, a step that made his geometric algebra more flexible – so flexible indeed that today we read xx as "x squared" without ever seeing a square in our mind's eye.

Explain the first part of this quotation.

2 Write a paragraph reflecting on the advantages of the law of homogeneity in terms of explaining algebra to beginners. In your answer make use of the diagrams in *The Arabs* unit for solving quadratic equations.

Practice exercises for this chapter are on page 155.

7 *Constructing algebraic solutions*

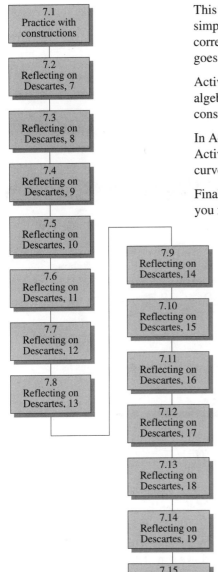
This chapter delves further into Descartes's work. After reminding you about some simple constructions, it shows how Descartes constructed the geometric lengths corresponding to algebraic expressions, using straight lines and circles. Descartes goes on to argue that other curves should also be allowed.

Activity 7.1 gives you practice at constructing geometric lengths corresponding to algebraic expressions. Activities 7.2 to 7.6 give a more systematic method for these constructions.

In Activities 7.7 to 7.9 you will see how Descartes considered other curves. In Activities 7.10 to 7.14, you will learn how he argued against the restriction of the curves used for the solution of construction problems to the straight line and circle.

Finally, in Activity 7.15, you will return to the problem of trisecting an angle, which you first considered in Chapter 6.

Work on the activities in sequence.

All the activities are fairly short, and you should try to work several of them in one sitting.

All the activities are suitable for small group working.

Introduction

Having introduced algebraic methods for solving geometric problems and also relaxed the law of homogeneity, Descartes had at his disposal a generally applicable method for analysing geometrical construction problems.

After an algebraic analysis, the next step in solving geometric problems is the translation of this analysis into a geometrical construction of the solution.

You have already seen some simple examples of these 'translations'. Here are some reminders.

Example 1

Let a, b, and c be line segments. Construct the line segment z such that

$$z = \frac{ab}{c}.$$

Descartes

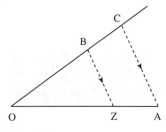

Figure 7.1

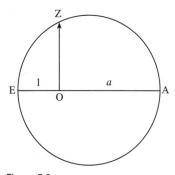

Figure 7.2

Solution

- Draw an arbitrary acute angle O, as shown in Figure 7.1.
- Mark off the distances b and c along one side of the angle such that OB $= b$ and OC $= c$, and mark off a along the other side so that OA $= a$.
- Draw BZ parallel to AC.
- Then OZ is the required line segment with length z.

Example 2

Let a be a line segment. Construct the line segment z such that $z = \sqrt{a}$.

Solution

- Take the length a and the length of the unit and draw them in line, as in Figure 7.2, so that OA $= a$ and OE $= 1$.
- Draw a circle with diameter AE.
- At O, draw a line perpendicular to AE. Call the point where this line intersects the circle, Z.
- Then OZ is the required line segment with length z.

In the next activity there are two constructions for practice. If you want to use a unit, take 1 cm as the unit.

Activity 7.1 Practice with constructions

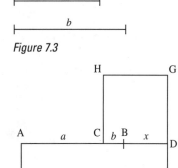

Figure 7.3

Figure 7.4

1 Let a and b be the two line segments shown in Figure 7.3. Construct the line segment z for $z = \sqrt{ab}$.

2 Look again at Van Schooten's problem in Activity 6.12. Figure 7.4 shows a straight line AB, with a point C on it. The problem was to find the point D on AB produced, such that the rectangle with sides AD and DB is equal (in area) to the square with side CD.

In Activity 6.12, you analysed this problem algebraically, and should have found that $x = \dfrac{b^2}{a-b}$, or $\dfrac{x}{b} = \dfrac{b}{a-b}$. Construct x using one of these expressions as a starting point.

Constructions requiring a circle

One of the 'rules of the game' applied to geometrical constructions from antiquity was that only a pair of compasses and a straight edge were used to execute the constructions. Descartes discusses this in *Book I*:

> 12   I shall therefore content myself with the statement that if the student, in solving these equations, does not fail to make use of division wherever possible, he will surely reach the simplest terms to which the problem can be reduced.

13   And if it can be solved by ordinary geometry, that is, by the use of straight lines and circles traced on a plane surface, when the last equation shall have been entirely solved there will remain at most only the square of an unknown quantity, equal to the product of its root by some known quantity, increased or diminished by some other quantity also known. Then this root or unknown line can easily be found.

Activity 7.2   Reflecting on Descartes, 7

1   Which algebraic expressions can, according to Descartes, be turned into geometric constructions with a pair of compasses and a straight edge?

2   Viète had shown that the algebraic formulation of the problems of duplicating the cube and trisecting the angle both led to cubic equations. What is the implication of paragraph 13 in the light of this?

Descartes continues his discussion:

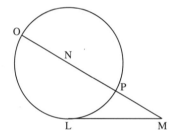

Figure 7.5

14   For example, if I have $z^2 = az + b^2$, I construct a right angle NLM with one side LM, equal to b, the square root of the known quantity b^2, and the other side, LN, equal to $\frac{1}{2}a$, that is, to half the other known quantity which was multiplied by z, which I supposed to be the unknown line. Then prolonging MN, the hypotenuse of this triangle to O, so that NO is equal to NL, the whole line OM is the required line z. This is expressed in the following way:

$$z = \tfrac{1}{2}a + \sqrt{\tfrac{1}{4}a^2 + b^2}$$

Figure 7.5 is a construction, using a circle and a line, of the unknown quantity z from the equation $z^2 = az + b^2$. The construction makes use of the property, $MP.MO = ML^2$, of the circle.

Activity 7.3   Reflecting on Descartes, 8

  1   Prove that for a circle $MP.MO = ML^2$ (Figure 7.5).

2   Show how you can construct Figure 7.5 from the given coefficients a and b. Use the result of question **1** to show that if $OM = z$ then $z^2 = az + b^2$.

3   The equation $z^2 = az + b^2$ has another solution, $z = \frac{1}{2}a - \sqrt{\frac{1}{4}a^2 + b^2}$. Why do you think Descartes did not consider this solution?

You can use Figure 7.5 to carry out another construction.

15   But if I have $y^2 = -ay + b^2$, where y is the quantity whose value is desired, I construct the same right triangle NLM, and on the hypotenuse MN lay off NP equal to NL, and the remainder PM is y, the desired root.

Thus I have

$$y = -\tfrac{1}{2}a + \sqrt{\tfrac{1}{4}a^2 + b^2}\,.$$

In the same way, if I had

$$x^4 = -ax^2 + b^2,$$

PM would be x^2 and I should have

$$x = \sqrt{-\tfrac{1}{2}a + \sqrt{\tfrac{1}{4}a^2 + b^2}}\,,$$

and so for other cases.

Activity 7.4   Reflecting on Descartes, 9

1 In the case of the equation $x^4 = -ax^2 + b^2$, PM will be equal to x^2 (Figure 7.5). Show how you can construct x from this result.

2 What can you say about the law of homogeneity in this case?

To continue:

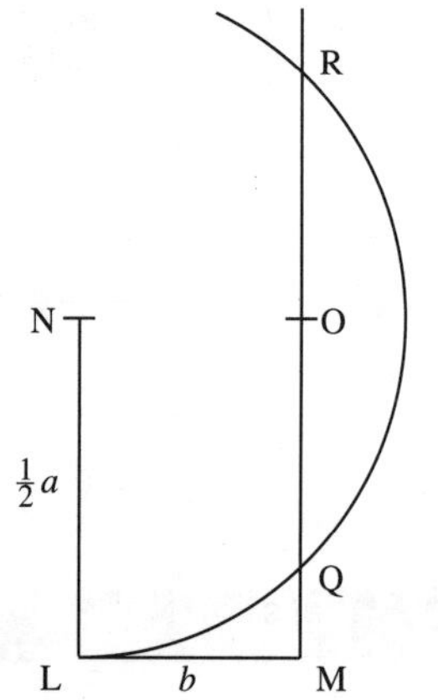

Figure 7.6

16 Finally, if I have $z^2 = az - b^2$, I make NL equal to $\tfrac{1}{2}a$ and LM equal to b as before; then, instead of joining the points M and N, I draw MQR parallel to LN, and with N as a centre describe a circle through L cutting MQR in the points Q and R; then z, the line sought, is either MQ or MR, for in this case it can be expressed in two ways, namely:

$$z = \tfrac{1}{2}a + \sqrt{\tfrac{1}{4}a^2 - b^2}\,,$$

and

$$z = \tfrac{1}{2}a - \sqrt{\tfrac{1}{4}a^2 - b^2}\,.$$

17 And if the circle described about N and passing through L neither cuts nor touches the line MQR, the equation has no root, so that we may say that the construction of the problem is impossible.

Activity 7.5   Reflecting on Descartes, 10

Figure 7.6 illustrates passages 16 and 17 above. The circle with centre N and radius $\tfrac{1}{2}a$ is drawn. NL is parallel to MR, and O is the mid-point of QR.

1 Show that NO is parallel to LM.

2 Express OQ in terms of a and b.

3 Express MQ and MR in terms of a and b.

4 Why does Descartes consider two solutions in this text?

5 In which case can a solution not be found? In which paragraph does Descartes describe this?

6   In the previous passages Descartes has considered the following equations.

$$z^2 + az - b^2 = 0$$

$$z^2 - az - b^2 = 0$$

$$z^2 - az + b^2 = 0$$

Why does Descartes not consider the equation $z^2 + az + b^2 = 0$?

Subsequently, Descartes compares his work to that of the mathematicians of antiquity.

> **18**   These same roots can be found by many other methods. I have given these very simple ones to show that it is possible to construct all the problems of ordinary geometry by doing no more than the little covered in the four figures that I have explained. This is one thing which I believe the ancient mathematicians did not observe, for otherwise they would not have put so much labour into writing so many books in which the very sequence of the propositions shows that they did not have a sure method of finding all, but rather gathered together those propositions on which they had happened by accident.

Activity 7.6   Reflecting on Descartes, 11

1   According to Descartes, what is missing from the work of previous mathematicians when solving geometrical construction problems?

In the remaining part of *Book I*, Descartes addresses a complex problem from antiquity, known as Pappus's problem. In treating Pappus's problem, Descartes drops a remark which is particularly apt as a conclusion for this section.

> **19**   Here I beg you to observe in passing that the considerations that forced ancient writers to use arithmetical terms in geometry, thus making it impossible for them to proceed beyond a point where they could see clearly the relation between the two subjects, caused much obscurity and embarrassment in their attempts at explanation.

Solutions which are curves

Descartes had previously said that a construction problem has a finite number of solutions if the algebraic formulation of the problem has as many equations as it has unknown quantities. In one of the previous quotations, paragraph 8 in Chapter 6, he touched briefly upon the problem that arises when there are more unknown quantities than there are equations.

Activity 7.7   Reflecting on Descartes, 12

1 Read paragraph 8 in Chapter 6 again. According to Descartes, what should be your strategy when there are more unknown quantities than there are equations?

In the case when there is one more unknown quantity than there are equations, you will find a solution curve.

> **20** Then, since there is always an infinite number of different points satisfying these requirements, it is also required to discover and trace the curve containing all such points.

At the beginning of Chapter 6 you met the example of the two points A and B and the set of points X for which $AX = 2BX$.

Activity 7.8   Reflecting on Descartes, 13

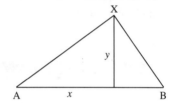

Figure 7.7

1 Show that the requirement $AX = 2BX$ gives the algebraic equation $3x^2 + 3y^2 - 8ax + 4a^2 = 0$. See Figure 7.7.

2 You already know that the solution set of this equation is a curve, in this case a circle. Find the centre and the radius of this circle from its equation.

In the previous example, the unknown line segments x and y are perpendicular to each other. You may suspect that, in this method of solution, a Cartesian x-y coordinate system is used. However, Descartes did not consciously use a perpendicular coordinate system. A coordinate system was, for Descartes, more a derived form of the sides of an angle, as used in the geometric constructions in Figures 7.5 and 7.6: an oblique coordinate system without 'negative axes'.

To conclude this section, here is a treatment of a simplified version of Pappus's problem, mentioned earlier.

Let l and m be two parallel lines a distance a apart. A third line n intersects l and m at right angles.

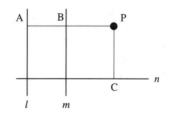

Figure 7.8

Find the point P such that 'the squared distance from P to l is equal to the product of the distance from P to m and the distance from P to n'.

This is equivalent to asking for the locus of points P such that $PA^2 = PB \times PC$.

Activity 7.9   Reflecting on Descartes, 14

1 Name the unknown quantities and formulate an algebraic equation for the above locus.

2 Use your graphics calculator to draw this solution curve.

On the nature of curves

In *La Géométrie*, Descartes pushed back some of the frontiers which had restricted mathematicians. You have already seen that the introduction of the unit and – more or less related to this – the relaxing of the law of homogeneity, enabled a fruitful interaction between geometry and algebra to take place. This cleared the way for a generally applicable method for solving geometrical construction problems.

However, yet another step was needed. What use is the method of analysis if you cannot then construct the solution? What about equations of third and higher degree? In general, the solutions of these equations cannot be constructed with a pair of compasses and a straight edge. But how can you construct them? Can you use other curves and circles?

Here, too, Descartes proved himself to be original and creative. Before he supported carrying out constructions with curves other than straight lines and circles, he discusses the nature of geometrical curves in *Book II*.

The book begins as follows.

> Geometry
>
> Book II
>
> On the Nature of Curved Lines
>
> 21   The ancients were familiar with the fact that the problems of geometry may be divided into three classes, namely, plane, solid, and linear problems. This is equivalent to saying that some problems require only circles and straight lines for their construction, while others require a conic section and still others require more complex curves. I am surprised, however, that they did not go further, and distinguish between different degrees of these more complex curves, nor do I see why they called the latter mechanical rather than geometrical.
>
> 22   If we say that they are called mechanical because some sort of instrument has to be used to describe them, then we must, to be consistent, reject circles and straight lines, since these cannot be described on paper without the use of compasses and a ruler, which may also be termed instruments. It is not because other instruments, being more complicated than the ruler and compasses, are therefore less accurate, for if this were so they would have to be excluded from mechanics, in which the accuracy of construction is even more important than in geometry. In the latter, the exactness of reasoning alone is sought, and this can surely be as thorough with reference to such lines as to simpler ones. I cannot believe, either, that it was because they did not wish to make more than two postulates, namely (1) a straight line can be drawn through any two points, and (2) about a given centre a circle can be described passing through a given point. In their treatment of the conic sections they did not hesitate to introduce the assumption that any given cone can be cut by a given plane. Now to treat all the curves which I mean to introduce here, only one additional assumption is necessary, namely, two or more lines can be moved, one upon the other, determining by their intersection other curves. This seems to me in no way more difficult.

> **23**   It is true that the conic sections were never freely received into ancient geometry, and I do not care to undertake to change names confirmed by usage; nevertheless, it seems very clear to me that if we make the usual assumption that geometry is precise and exact, while mechanics is not; and if we think of geometry as the science which furnishes a general knowledge of the measurement of all bodies, then we have no more right to exclude the more complex curves than the simpler ones, provided they can be conceived of as described by a continuous motion or by several successive motions, each motion being completely determined by those which precede; ... .

Descartes starts by mentioning three kinds of geometrical construction problems: plane, solid and linear. The terms 'plane', 'solid' and 'linear' relate to the curves that are needed to construct the solutions. This three-way classification dates from antiquity (it is mentioned for the first time in the work of Pappus) and calls for some explanation.

In Chapter 6 you carried out a number of constructions with lines and circles as 'curves'. These curves originated in antiquity in the plane (two-dimensional space) and are therefore called plane curves. The geometrical construction problems solved with these curves are consequently called plane problems.

The conic sections – the ellipse, the parabola and the hyperbola – were known and used by the mathematicians of antiquity. The conic sections were found as the 'intersection curves' of a cone and a plane, shown in Figure 7.9. An elaborate theory on this subject can be found in the famous work *Conica* from the Greek mathematician Apollonius of Perga (262–190 BC). It is remarkable that almost all properties of these curves, as we know them now from the analytical geometry, can be found in Apollonius's work.

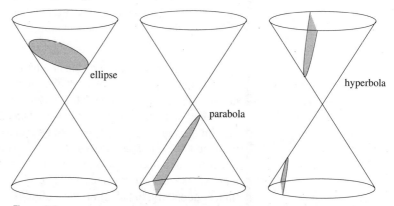

Figure 7.9

As these conic sections originated in a solid three-dimensional figure, the cone, they were called solid curves. In Greek antiquity, constructions with these curves were described, but the search for constructions in two dimensions that could replace them was continued, since these were preferred. It was only in the 19th century that it became clear that a number of geometrical construction problems including the trisection of the angle could be solved only with solid curves.

Finally, there were other, more complicated, curves, such as the spiral, the quadratrix, the cissoid and the conchoid. These curves were sometimes used to carry out a construction, more as a curiosity than as an accepted geometrical construction. For example, the problem of doubling a cube could be solved with a cissoid as carried out by Diocles in about 180 BC. These curves were described as linear curves in the antiquity.

Look back at paragraphs 21 to 23.

1 Descartes objects to the division of geometrical problems into three classes by the mathematicians in antiquity. What is his opinion on the third class, the linear problems?

2 How did the mathematicians of antiquity discriminate between 'geometrical curves' and 'mechanical curves'? What is Descartes's opinion on this?

3 Descartes mentions two postulates from Euclid's *Elements*. Which are they?

4 Why does Descartes discuss these postulates here and not in another part of *La Géométrie*?

5 In paragraph 22 Descartes suggests that if straight edge and compasses are acceptable instruments for the construction of curves then there might be other instruments that are equally acceptable. Which conditions must be met, according to Descartes, on how these instruments work?

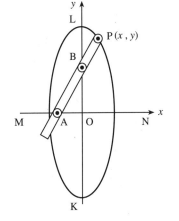

Figure 7.10

Instruments for the construction of curves

Figure 7.10 shows how you can make an instrument to draw an ellipse.

The pin at A moves 'horizontally' along the slot MON which lies along the *x*-axis while the pin at B moves 'vertically' in the slot KOL which lies along the *y*-axis. The pointer P is a point along the arm AB so that the lengths AP and BP are fixed. As A and B move, so P traces a path.

Let the point P have coordinates (x, y), and suppose that $PA = a$, $PB = b$ and $AB = c$ so that $a = b + c$.

1 Show that: $y = \dfrac{a}{c} y_B$ and $x = -\dfrac{b}{c} x_A$.

2 Show that the point P describes an ellipse.

3 Descartes accepts a curve under the condition that it has originated from a movement that can be linked in a direct and clear way to a straight or a circular movement. Does the construction with this instrument meet this requirement?

Figure 7.11 shows a geometrical instrument described in *De Organica Conicarum Sectionum in Plano Descriptione Tractatus* (Leiden, 1646) by Frans van Schooten.

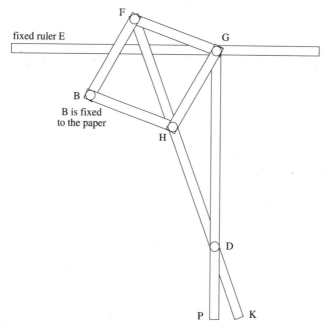

Figure 7.11

The ruler E is fixed firmly on the paper. The strips BF, BH, GF and GH are equal in length. At the point G, which is the 'hinge' of the strips FG, GH and GP, is a pin that can be moved along a slot in the fixed ruler E. During this movement, the strip GP remains perpendicular to the ruler E. The 'hinge' B of the strips FB and BH is fixed to the paper by using a pin. The point H can move freely along the strip FK (by moving a pin in a slot). Finally, at D, the point of intersection of the strips GP and FK which can move in a slot in the strip GP, there is a pen which traces out a curve as G moves along the ruler.

Activity 7.12   Reflecting on Descartes, 17

1 By making a good choice for an *x-y* coordinate system, show algebraically that the curve produced by the instrument in Figure 7.11 is part of a parabola.

2 Alternatively, every point of a parabola has the same distance to a fixed point, called the focus, and a fixed line called the directrix. Prove that the point D has this property. Use similar triangles to prove that $BD = GD$.

3 Does this instrument fulfil the conditions that Descartes sets for the movements of such instruments?

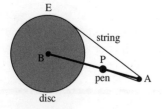

Figure 7.12

In 1650, Christiaan Huygens (1629–1695), a student of Frans van Schooten, and thus indirectly of Descartes, made a sketch of a mechanism to construct a spiral. This mechanism is shown in Figure 7.12.

The point B is the centre of a fixed circular disc, and E is a fixed point on its circumference. AB is a ruler which can turn about B. A string is fixed to the disc at E, passes round the pulley at A and is then attached to a pen P which can slide up and down the ruler AB. As A turns, the string winds round the disc, round the pulley at A and the pen P moves up and down the ruler, drawing a spiral shape.

Even though the movements of the ruler and pen are directly coupled, Descartes did not think that this was an acceptable combination of movements. The relation between the two movements involves π, the ratio between the circumference and the diameter of a circle. Descartes thought that he did not know π with sufficient accuracy.

In *Book II*, Descartes gives an example of an instrument to reinforce his arguments in favour of the extension of the number of 'construction curves'. See Figure 7.13.

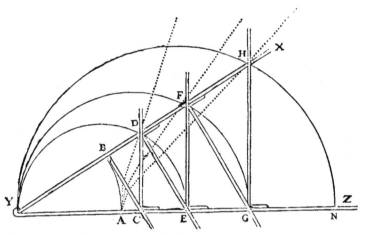

Figure 7.13

24　Consider the lines AB, AD, AF, and so forth, which we may suppose to be described by means of the instrument YZ. This instrument consists of several rulers hinged together in such a way that YZ being placed along the line AN the angle XYZ can be increased or decreased in size, and when its sides are together the points B, C, D, E, F, G, H, all coincide with A; but as the size of the angle is increased, the ruler BC, fastened at right angles to XY at the point B, pushes toward Z the ruler CD which slides along YZ always at right angles. In like manner, CD pushes DE which slides along YX always parallel to BC; DE pushes EF; EF pushes FG; FG pushes GH, and so on. Thus we may imagine an infinity of rulers, each pushing another, half of them making equal angles with YX and the rest with YZ.

25　Now as the angle XYZ is increased the point B describes the curve AB, which is a circle; while the intersections of the other rulers, namely, the points D, F, H describe other curves, AD, AF, AH, of which the latter are more complex than the first and this more complex than the circle. Nevertheless I see no reason why the description of the first cannot be conceived as clearly and distinctly as that of the circle, or at least as that of the conic sections; or why that of the second, third, or any other that can thus be described, cannot be as clearly conceived of as the first; and therefore I can see no reason why they should not be used in the same way in the solution of geometric problems.

It is clear that the points B, D, F and H describe curves of different complexity. Point B describes a circle, which is acceptable as a 'construction curve'. The movements of points D, F and H, however, are directly and clearly coupled to the movement of point B, and, as Descartes argues, why would these curves not be acceptable as well?

1 Let YA $= a$, YC $= x$ and CD $= y$ in Figure 7.13. Show that the curve AD can be described by the equation $x^4 = a^2\left(x^2 + y^2\right)$.

Acceptable curves

The previous section shows that Descartes did not want to restrict the set of acceptable construction curves to the line and the circle. He wanted to expand this set under one condition: the curve must be described by a continuous movement or by an acceptable sequence of continuous movements, in which every movement is completely determined by its predecessor. The line and the circle remain the starting points; they act as the original 'predecessor' of all curves.

Descartes knew that this condition does not directly lead to an immediately recognisable set of curves. Read what he writes about this.

> **26**   I could give here several other ways of tracing and conceiving a series of curved lines, each curve more complex than any preceding one, but I think the best way to group together all such curves and then classify them in order, is by recognising the fact that all points of those curves which we may call "geometric", that is, those which admit of precise and exact measurement, must bear a definite relation to all points of a straight line, and that this relation must be expressed by means of a single equation. If this equation contains no term of higher degree than the rectangle of two unknown quantities, or the square of one, the curve belongs to the first and simplest class, which contains only the circle, the parabola, the hyperbola, and the ellipse; but when the equation contains one or more terms of the third or fourth degree in one or both of the two unknown quantities (for it requires two unknown quantities to express the relation between two points) the curve belongs to the second class; and if the equation contains a term of the fifth or sixth degree in either or both of the unknown quantities the curve belongs to the third class, and so on indefinitely.

According to many this part of the text initiates the birth of analytical geometry: all points of a curve are unambiguously related to points on a straight line, and this relation can be described with one single equation with two unknown quantities.

1 Descartes mentions the circle, the parabola, the hyperbola and the ellipse as the representatives of the simplest class of curves (those with no terms higher than the second degree). Did he forget one? If so, which one?

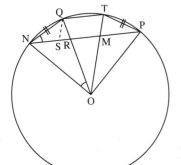

Figure 7.14

To conclude this chapter, here is an example of a geometrical construction corresponding to an equation of the third degree: the trisection of an angle, which you saw at the end of Chapter 6. Descartes shows this construction in *Book III*.

First his analysis:

> **27** Again, let it be required to divide the angle NOP, or rather, the circular arc NQTP, into three equal parts. Let NO = 1 be the radius of the circle, NP = q be the chord subtending the given arc, and NQ = z be the chord subtending one-third of that arc; then the equation is $z^3 = 3z - q$. For, drawing NQ, OQ and OT, and drawing QS parallel to TO, it is obvious that NO is to NQ as NQ is to QR as QR is to RS. Since NO = 1 and NQ = z, then QR = z^2 and RS = z^3; and since NP or q lacks only RS or z^3 of being three times NQ or z, we have $q = 3z - z^3$ or $z^3 = 3z - q$.

In a way similar to that in Activity 6.13, Descartes arrives at the equation $z^3 = 3z - q$.

Descartes uses a parabola and a circle in his construction of the length z. The method is similar to that of Omar Khayyam, as described in Chapter 5. It is shown in modern notation in Figure 7.15.

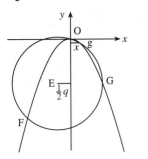

Figure 7.15

The figure shows the graph of the parabola $y = -x^2$ and the circle centre E $\left(-\frac{1}{2}q, -2\right)$ passing through the origin O. The circle and the parabola intersect in four points: O, g, G and F. The x-value of g is the length x that can be used in the trisection of the angle.

1 Find the equation of the circle.

2 Show that for the x-values of g, G, and F, the equation $z^3 = 3z - q$ holds.

Reflecting on Chapter 7

What you should know

- that Descartes made a major contribution to the development of mathematics by relaxing the rule that only straight edges and circles could be used in geometric constructions
- the contribution made by van Schooten to Descartes's work

- how to construct solutions to the following:

$$z^2 = az + b^2$$
$$y^2 = -ay + b^2$$
$$z^2 = az - b^2$$
$$z^3 = 3z - q$$

Preparing for your next review

- Reflect on the 'What you should know list' for this chapter. Be ready for a discussion on any of the points.
- Answer the following check questions.

1 Make sure that you can explain the construction for solving the equation $z^2 = az + b^2$.

2 Give an example of a problem which cannot be solved by using a straight edge and compasses only, but can be solved by using other curves.

3 What other curves are allowed by Descartes?

Practice exercises for this chapter are on page 155.

8 An overview of La Géométrie

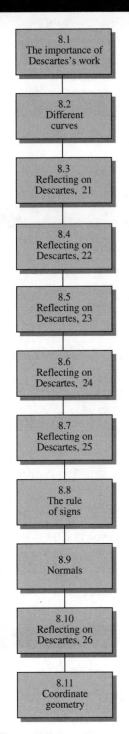

This chapter gives an overview of Descartes's work on curves. It then goes on to show how this led Descartes into some work on the solution of equations, the Descartes's rule of signs, which enables you to gain some information about whether the roots are positive or negative.

The first of the final two sections deals with finding normals to curves. The second compares the mathematical work of Descartes and his contemporary Frenchman, Fermat. The work of Fermat and Descartes together forms the starting point of analytical geometry.

Activity 8.1 is concerned with the general importance of Descartes's work that you have seen so far. Then in Activities 8.2 to 8.4 you will learn about his attempts to classify curves.

Activities 8.5 to 8.8 are about the work that Descartes did on finding the number of roots of an algebraic equation from the number of sign changes of its coefficients. Activity 8.9 shows how Descartes found the equations of normals to conics.

Activities 8.10 and 8.11 compare the work of Descartes with that of his contemporary, Fermat.

Work through the activities in sequence.

Descartes's *La Géométrie* is a revolutionary work in the history of mathematics.

Its purpose – finding a generally applicable method for geometrical constructions – was not new. The revolutionary aspects of the work lie primarily in the new methods he pioneered.

The work itself, on geometric constructions, really only supported and illustrated Descartes's philosophical thinking, even though it changed the mathematics of the time completely. However, its effect was that geometrical constructions, which had been a subject of study for centuries, received less, rather than more, attention soon after *La Géométrie* was published.

Here are the four major new features of Descartes's method.
● Using algebra to analyse geometrical construction problems.

Reduce the geometrical problem to a set of given and required line segments. Give letters to these line segments and translate the problem into algebraic relations between these letters. This removes the need to draw every new construction step by step. Instead, you can find the final construction by manipulating algebraic equations.

- Introducing a unit and relaxing the law of homogeneity.

By choosing a line segment as the unit, you can consider areas, volumes, and all other quantities with a dimension different from that of a line segment, as line segments. In this way algebra is released from its geometric constraints and becomes a powerful tool.

- Translating algebraic equations into corresponding geometrical constructions.

If the algebraic equation has been reduced to its simplest form, the final construction of the unknown line segment follows immediately. The line and the circle are no longer the only construction curves. Other curves are now equally acceptable, provided that they can be described by a sequence of continuous movements which are in the end directly and clearly coupled to a straight or to a circular movement. In the final construction, the most suitable curve must be used.

- A modern algebraic notation.

Apart from a few exceptions, Descartes's notation is the same as today's.

Activity 8.1   The importance of Descartes's work

Be prepared to bring your answer to this activity to your next review.

1   Which of the features claimed as revolutionary has impressed you most?

Write your answer with reasons in five sentences at most. Include:
- breaking with old traditions
- the importance for today's mathematics.

La Géométrie contains even more revolutionary ideas. Some of them will be mentioned in this chapter, if only briefly.

It is remarkable that the third point above – translating algebraic equations into geometric constructions – receives the most attention in *La Géométrie*. Descartes did much new work in this field, which was at the frontiers of research.

Here are some questions which arise from this work.
- Which curves, besides the circle and the line, are acceptable as construction curves?
- What is the simplest form of an algebraic equation and how do you find it?
- What curves are most suitable for a given geometrical construction?

A paragraph of *La Géométrie*, quoted in paragraph 26 of Chapter 7 of this book, shows that Descartes had started looking for a kind of classification of curves.

Activity 8.2   Different curves

Read once again paragraph 26 from *La Géométrie*.

1   According to Descartes, what property must be the distinguishing factor between the different classes of curves?

Descartes used a novel way of classifying curves to find which one he needed to solve a geometrical construction problem. At the same time he formulated a classification of construction problems.

In *Book II*, Descartes wrote extensively about curves but he did not derive equations of curves. Equations of curves, however, were very new and were not the goal of Descartes's study but merely a means to come to a classification. Yet, it was clear to him that equations contain a great deal of information about curves.

> 28 When the relation between all points of a curve and all points of a straight line is known, in the way I have already explained, it is easy to find the relation between the points of the curve and all other given points and lines; and from these relations to find its diameters, axes, centre and other lines or points which have especial significance for this curve, and thence to conceive various ways of describing the curve, and to choose the easiest.

> 29 By this method alone it is then possible to find out all that can be determined about the magnitude of their areas, and there is no need for further explanation from me.

Activity 8.3 Reflecting on Descartes, 21

1 If you want to characterise a curve today, giving an equation will do. To Descartes this was not (yet) acceptable. From which sentence in paragraphs 28 and 29 can you conclude this?

In paragraph 30, Descartes shows for the first time a method for determining the normal to a curve. He regarded the normal as very useful for discovering a number of properties of curves.

> 30 Finally, all other properties of curves depend only on the angles which these curves make with other lines. But the angle formed by two intersecting curves can be as easily measured as the angle between two straight lines, provided that a straight line can be drawn making right angles with one of these curves at its point of intersection with the other. This is my reason for believing that I shall have given here a sufficient introduction to the study of curves when I have given a general method of drawing a straight line making right angles with a curve at an arbitrarily chosen point upon it. And I dare say that this is not only the most useful and most general problem in geometry that I know, but even that I have ever desired to know.

Activity 8.4 Reflecting on Descartes, 22

1 From paragraph 30, it is clear that an important property of curves can be determined with the help of normals. Which property is meant?

In the next section you will go deeper into the method for finding the normal to a curve and see how this process is a forerunner of modern differential calculus.

For now, take a look at another part of the text, the beginning of *Book III*.

On the Construction of Solid and Supersolid Problems

31   While it is true that every curve that can be described by a continuous motion should be recognised in geometry, this does not mean that we should use at random the first one that we meet in the construction of a given problem. We should always choose with care the simplest curve that can be used in the solution of a problem, but it should be noted that the simplest means not merely the one most easily described, nor the one that leads to the easiest demonstration or construction of the problem, but rather the one of the simplest class that can be used to determine the required quantity.

32   … On the other hand, it would be a blunder to try vainly to construct a problem by means of a class of lines simpler than its nature allows.

Descartes says that, for every construction, you have to use a curve from a class that is as simple as possible. But take care not to choose too simple a class. If you want to construct the solution of a second-order equation, do not take a parabola or an ellipse. Take a circle and a line. However, the construction cannot be performed only with lines.

He goes on to say that, before you start, you should ensure that the algebraic equation is in its simplest form. Descartes writes in detail about this in *Book III*.

He starts by saying:

33   Before giving the rules for the avoidance of both these errors, some general statements must be made concerning the nature of equations. An equation consists of several terms, some known and some unknown, some of which are together equal to the rest; or rather, all of which taken together are equal to nothing; for this is often the best form to consider.

Activity 8.5   Reflecting on Descartes, 23

1   According to Descartes, what is the best form in which to write
$x^3 = 3x^2 + 2x - 6$?

In *Book III*, he explores the theory of algebraic equations, with the aim of simplifying an equation and if possible lowering its order. This results in several ideas that were remarkably original. These can be seen in paragraphs 34 and 35.

34   Every equation can have as many distinct roots (values of the unknown quantity) as the number of dimensions of the unknown quantity in the equation. Suppose, for example, $x = 2$ or $x - 2 = 0$, and again, $x = 3$ or $x - 3 = 0$. Multiplying together the two equations $x - 2 = 0$ and $x - 3 = 0$, we have $x^2 - 5x + 6 = 0$, or $x^2 = 5x - 6$. This is an equation in which x has the value 2 and at the same time x has the value 3. If we next

make $x - 4 = 0$ and multiply this by $x^2 - 5x + 6 = 0$, we have $x^3 - 9x^2 + 26x - 24 = 0$ another equation, in which x, having three dimensions, has also three values, namely, 2, 3, and 4.

35   It often happens, however, that some of the roots are false or less than nothing. Thus, if we suppose x to represent the defect of a quantity 5, we have $x + 5 = 0$ which, multiplied by $x^3 - 9x^2 + 26x - 24 = 0$, yields $x^4 - 4x^3 - 19x^2 + 106x - 120 = 0$, an equation having four roots, namely three true roots, 2, 3, and 4, and one false root, 5.

Activity 8.6   Reflecting on Descartes, 24

1   At that time, when the geometrical interpretation of algebra was still common, it was very unusual to allow negative solutions of equations, as Descartes does. How does he refer to these negative solutions?

2   What is the meaning of 'the defect of a quantity 5,' in paragraph 35?

Apart from negative solutions, irrational numbers appear as solutions and coefficients of equations.

Here is a rule which has gone down in history as Descartes's rule of signs.

36   ... We can determine also the number of true and false roots that any equation can have, as follows: An equation can have as many true roots as it contains changes of sign, from + to − or from − to +; and as many false roots as the number of times two + signs or two − signs are found in succession.

37   Thus, in the last equation, since $+x^4$ is followed by $-4x^3$, giving a change of sign from + to −, and $-19x^2$ is followed by $+106x$ and $+106x$ by -120, giving two more changes, we know there are three true roots; and since $-4x^3$ is followed by $-19x^2$ there is one false root.

Activity 8.7   Reflecting on Descartes, 25

1   How many positive solutions does the sign rule suggest for the equation $x^2 - x - 6 = 0$? And how many for $x^3 = 3x^2 + 4x - 12$?

2   How many positive solutions does the sign rule predict for the equation $x^2 - x + 6 = 0$?

Descartes

Today we call these numbers complex numbers. The current notation for $\frac{1}{2}+\frac{1}{2}\sqrt{-23}$ is $\frac{1}{2}+\frac{1}{2}\sqrt{23}\,i$. The character i indicates $\sqrt{-1}$. A complex number is thus composed of a real part, here $\frac{1}{2}$, and an imaginary part, $\frac{1}{2}\sqrt{23}$.

Apparently, Descartes also considers the solutions of $x^2-x+6=0$ to be positive solutions: $\frac{1}{2}+\frac{1}{2}\sqrt{-23}$ and $\frac{1}{2}-\frac{1}{2}\sqrt{-23}$.

In *La Géométrie* these so called imaginary numbers are mentioned explicitly.

> 38 Neither the true nor the false roots are always real; sometimes they are imaginary; that is, while we can always conceive of as many roots for each equation as I have already assigned, yet there is not always a definite quantity corresponding to each root so conceived of. Thus, while we may conceive of the equation $x^3-6x^2+13x-10=0$ as having three roots, yet there is only one real root, 2, while the other two, however we may increase, diminish, or multiply them in accordance with the rules just laid down, remain always imaginary.

From Descartes's statement that the number of solutions of an equation is equal to its order, you can conclude that he accepted imaginary solutions of equations.

Activity 8.8   The rule of signs

1   The equation x^4-5x^2+4 can be factorised into $(x-1)(x+1)(x-2)(x+2)$. Can the sign rule be applied to $x^4-5x^2+4=0$?

2   In the previous chapters, the trisection of an angle resulted in the equation $z^3=3z-q$, with $q>0$. How many positive solutions has this equation?

3   The construction of solutions of the equation $z^3=3z-q$ was carried out in Chapter 7, Activity 7.15. The solutions are the x-values of g, G and F. Which of these three solutions is negative?

Descartes's ideas about equations provides a tool for reducing algebraic equations to their simplest form. This nearly completes his attempts to solve geometrical construction problems in a generally applicable method. He ends his book with examples of constructions, among which is the trisection of an angle.

All in all, *La Géométrie* is a book full of revolutionary ideas, even though it was intended to tackle a classical goal, namely, solving geometrical construction problems. The book had a great effect on mathematics: it caused the birth of analytical geometry and the appearance of differential calculus. Descartes's wish, that future generations would appreciate his work, has come true.

> 39   But it is not my purpose to write a large book. I am trying rather to include much in a few words, as will perhaps be inferred from what I have done, if it is considered that, while reducing to a single construction all the problems of one class, I have at the same time given a method of transforming them into an infinity of others, and thus of solving each in an infinite number of ways; that, furthermore, having constructed all plane problems by the cutting of a circle by a straight line, and all solid problems by the cutting of a circle by a parabola; and, finally, all that are but one degree more complex by cutting a circle by a curve but one degree higher than the parabola, it is only necessary to follow the same general method

> to construct all problems, more and more complex, ad infinitum; for in the case of a mathematical progression, whenever the first two or three terms are given, it is easy to find the rest.

> 40   I hope that posterity will judge me kindly, not only as to the things which I have explained, but also as to those which I have intentionally omitted so as to leave to others the pleasure of discovery.

The method of normals

The full title is *Discours de la méthode pour bien conduire sa raison et chercher la verite dans les sciences.*

Descartes wrote *La Géométrie* as an appendix to his philosophical work *Discours*. Another appendix is *Dioptrique*, a treatment on geometrical optics. To draw the light path made by an incoming ray on a lens, it is essential to construct the normal to a curve. In *La Géométrie* Descartes shows a method for constructing normals.

Figure 8.1 shows part of an ellipse, as it appeared in *La Géométrie*.

Here is Descartes's solution to finding the normal to a curve, when the curve is an ellipse. The method is based on the following idea.

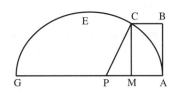

Figure 8.1

> 41   … observe that if the point P fulfils the required conditions, the circle about P as centre and passing through the point C will touch but not cut the curve CE; but if this point P be ever so little nearer to or farther from A than it should be, this circle must cut the curve not only at C but also in another point.

In short, point P, the centre of a circle with radius PC, has a fixed place on GA, as the circle intersects with the ellipse in C. If PC were not perpendicular to the ellipse, the circle would intersect with the ellipse at another point as well.

Descartes names various known and unknown quantities.

> 42   Suppose the problem solved, and let the required line be CP. Produce CP to meet the straight line GA, to whose points the points of CE are to be related. Then, let $MA = CB = y$; and $CM = BA = x$. An equation must be found expressing the relation between x and y. I let $PC = s$, $PA = v$, whence $PM = v - y$.

The ellipse and the circle can be represented as two equations in x and y (the parameters of the ellipse are known, so its equation can be found). The problem can now be solved algebraically. You can eliminate x or y from the equations, leaving one equation with one unknown quantity.

Suppose that C is the point of contact of the ellipse and the circle. What happens then? Or, put in another way, what happens if the circle and the ellipse have two coincident points of intersection?

> 43   … and when the points coincide, the roots are exactly equal, that is to say, the circle through C will touch the curve CE at the point C without cutting it.

In that case the equation has a double solution.

In Activity 8.9, you will retrace this solution with a concrete example.

Activity 8.9   Normals

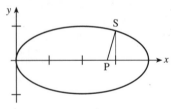

Figure 8.2

Figure 8.2 shows an ellipse with its major axis along the *x*-axis. The length of the major axis is 4, and the length of the minor axis is 2.

1 Check that the equation of this ellipse is $x^2 - 4x + 4y^2 = 0$.

The point $P(p,0)$ is on the *x*-axis, and the *x*-coordinate of the point S on the ellipse, is 3. The circle with centre P and radius PS touches the ellipse.

2 Show that
a the equation of the circle is $y^2 = -x^2 + 2px + 9\frac{3}{4} - 6p$
b the *x*-coordinate of the point of contact S of the circle and the ellipse is $x^2 + \frac{1}{3}(4 - 8p)x + 8p - 13 = 0$
c this implies that $\frac{1}{3}(4 - 8p) = -6$ and $8p - 13 = 9$.

3 Find the equation of the normal at S to the ellipse.

4 Find in the same way the equation of the normal at $(4,3)$ to the parabola $y = \frac{1}{2}(x - 2)^2 + 1$.

Descartes, Fermat and analytical geometry

Descartes and his *La Géométrie* are often closely associated with the development of analytical or coordinate geometry, but this is only partly valid. *La Géométrie* is not a book on analytical geometry, but, as you have seen, it acted as a starting point for the study of analytical geometry. But Descartes was not the only one who did the groundwork; among others who contributed was his contemporary, Pierre Fermat.

Fermat's mathematical work was partly in the same field as that of Descartes. There are some similarities, but also a number of clear differences. In this section you will briefly compare some aspects of their work.

Analytical geometry is the use of coordinate systems and algebraic methods in geometry. In a coordinate system, a point is represented as a set of numbers (coordinates) and a curve is represented by an equation for a specific set of points. So the geometrical properties of curves and figures can be investigated with the help of algebra.

Activity 8.10   Reflecting on Descartes, 26

1 Describe at least one aspect of *La Géométrie* closely related to the above definition of analytical geometry.

2 Describe one aspect that does not relate to this definition.

Mathematics was not one of Pierre Fermat's major activities, in contrast to Descartes. Fermat (1601–1665) was a lawyer who worked in the Parliament of the city of Toulouse. For him, mathematics was a hobby.

Fermat as well as Descartes did revolutionary work uniting geometry and algebra. His work was of less influence than Descartes's because it was not published until 1679, after Fermat's death. His book *Ad Locos Planos et Solidos Isagoge*, in which for the first time a kind of introduction to analytical geometry can be found, was probably written in 1629, that is, well before *La Géométrie*. In his book, he discusses systematically simple equations with two unknown quantities and draws the corresponding curves, lines, circles, ellipses, parabolas and hyperbolas.

Fermat goes much further than Descartes with the equations of curves. His systematic approach to curves and equations is clearer, but this can be explained by the fact that his intention with *Isagoge* was very different from Descartes's intention with *La Géométrie*.

La Géométrie	*Ad Locos Planos et Solidos Isagoge*
• Finding solutions of geometrical construction problems with the help of algebra.	• Studying curves with the help of algebra.
• The emphasis is on finding lengths.	• The emphasis is on the conic sections of Apollonius.
Seen in an algebraic light this means that **the emphasis is on equations with one unknown quantity**.	Seen in an algebraic light this means that **the emphasis is on equations with two unknown quantities**.

In one respect, Fermat was clearly more traditional than Descartes; he applied algebra as Viète meant algebra to be applied. This was the main reason that his *Isagoge* was already considered old-fashioned when it was published.

Here is an example that shows how Fermat considered the hyperbola. He uses vowels for the unknown quantities, A and E in this example, and consonants for known quantities, Z in this example. Fermat shows the curves in Figure 8.3. Point I is characterised by NP ($= A$) and PI ($= E$).

Fermat wrote: '$A.E = Z$ pl. The rectangles NPI and NMO are equal. Point I thus describes a hyperbola with asymptotes NM and NR.' (Today we write $xy = c$ instead of $A.E = Z$ pl.)

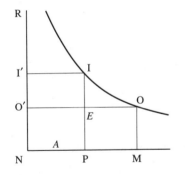

Figure 8.3

Activity 8.11   Coordinate geometry

1   What is the meaning of the addition 'pl' in Z pl?

2   a   Comment on this example. Notice that coordinate axes, quadrants, positive and negative numbers are used and that the curves are complete.
b   Compare this example with both Descartes's and today's mathematics.

Fermat wrote several mathematical works that were published after his death. However, some of his work was widely known earlier, through his correspondence, including a method for finding maxima and minima of curves which is so close to the current method that Fermat is considered by many to be the founder of differential calculus. His method for finding areas under specific curves was also far ahead of his time. He was a brilliant amateur mathematician.

Descartes's only mathematical work was *La Géométrie*. He did correspond with other mathematicians, among them Fermat.

What has happened subsequently to analytical geometry since Descartes and Fermat? Very soon, other mathematicians adopted the Cartesian approach and systematic research into curves flourished. Algebra, without the law of homogeneity, became a powerful tool for analysing geometry.

Once differential calculus had been developed, by the end of the 17th century, virtually all the problems of classical geometry were tackled and solved. Mathematicians of later date, among them Euler, also contributed to the intellectual achievements now known as analytical geometry.

Reflecting on Chapter 8

What you should know

- how the work of Descartes compares with that of Pierre Fermat
- evidence of Descartes's approach to negative and imaginary solutions of equations
- how Descartes constructed a normal to a given curve
- the relationship between algebra and geometry in *La Géométrie*.

Preparing for your next review

- Bring your answer to Activity 8.1 to the review together with half a page in your own words describing the revolutionary features of *La Géométrie*.
- Answer the following check questions.

1 Prepare a presentation explaining how the work of Descartes was influenced by van Schooten, Viète and Fermat.

2 Use Descartes's method to find the equation of the normal to $y^2 = 4x$ at $(1, 2)$.

Practice exercises for this chapter are on page 155.

5

Calculus

Introduction

Chapter 9
The beginnings
of calculus

In this unit, you will learn how, before the introduction of calculus, mathematicians found tangents and areas enclosed by some of the simpler curves. You will then see what was distinctive about the work of Newton and Leibniz.

This unit is designed to take about 10 hours of your learning time. About half of this time will be outside the classroom.

There are summaries and further practice exercises in Chapter 12.

Mathematical knowledge assumed

- knowledge of calculus will help you with this unit
- you should have studied the work of Archimedes in Activity 4.7.

9 The beginnings of calculus

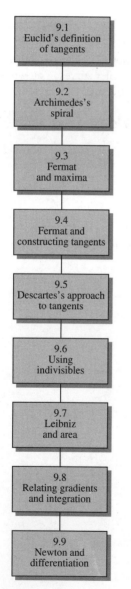

| 9.1 |
| Euclid's definition |
| of tangents |

| 9.2 |
| Archimedes's |
| spiral |

| 9.3 |
| Fermat |
| and maxima |

| 9.4 |
| Fermat and |
| constructing tangents |

| 9.5 |
| Descartes's approach |
| to tangents |

| 9.6 |
| Using |
| indivisibles |

| 9.7 |
| Leibniz |
| and area |

| 9.8 |
| Relating gradients |
| and integration |

| 9.9 |
| Newton and |
| differentiation |

Calculus was developed to solve two major problems with which many mathematicians had worked for centuries. One was the problem of finding extreme values of curves – the problem now called finding maxima and minima. The other was the problem of quadrature, or finding a systematic way to calculate the areas enclosed by curves.

Until Leibniz and Newton applied themselves to the problems, several different geometrical methods existed for solving these problems in specific cases. There was no general method which would work for any curve.

The chapter begins with some of these geometrical methods, so that you can see how difficult they are and realise why there was such a need for a general method. It goes on to show something of the pioneering work of Fermat, Descartes, Cavalieri, Leibniz and Newton as they developed their ideas.

Activities 9.1 and 9.2 show how Euclid and Archimedes drew tangents to two very special curves. In Activities 9.3 to 9.5 you see how Fermat and then Descartes approached the idea of drawing a tangent to a curve. Activities 9.6 and 9.7 show the beginnings of the modern ideas and notations for integrals.

Activity 9.8 shows how Leibniz thought of the relation between differentiation and integration, and Activity 9.9 shows how Newton developed a system for differentiation.

All the activities are suitable for working in a small group.

Work through the activities in sequence.

Early methods for finding tangents

To find an extreme value, you have to find a tangent to a curve. For the Greeks, this was a problem of pure geometry. Later, in the 17th century, it became important for the science of optics studied by Fermat, Descartes, Huygens and Newton, and in the science of motion studied by Newton and Galileo.

The work that the Greeks were able to do was limited by three factors:
● their unsatisfactory ideas of angle – they tried to include angles in which at least one of the sides was curved (see Figure 9.1)

Figure 9.1

- a definition of tangent which they could use only for simple curves
- a limited number of fairly simple curves that they could study.

Their method for proving that a line was a tangent to a curve involved:
- specifying a construction for a tangent line at any given point P on the curve
- proving that the line so constructed does not meet the curve anywhere else.

Activity 9.1   Euclid's definition of tangents

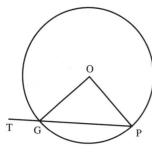

Figure 9.2

Euclid, in *Book III*, proposition 16, proved that the line PT in Figure 9.2 is a tangent to the circle at P if angle OPT is a right angle.

> The straight line drawn at right angles to the diameter of a circle at its extremity will fall outside the circle, and into the space between the straight line and the circumference another straight line cannot be interposed; further the angle of the semicircle is greater, and the remaining angle less, than any acute rectilineal angle.

1   Complete Euclid's argument by filling in the gaps.

Suppose that PT is a tangent to the circle. Then suppose that PT meets the circle at another point G.

Then OG = ...

This implies that angle OGP = ...

This means that the triangle ...... contains ...... right angles, which is impossible.

> This is an example of proof by contradiction.

Therefore the line PT cannot cut the circle anywhere else, and is proved to be a tangent, according to the definition prevailing in Euclid's time.

Activity 9.2   Archimedes's spiral

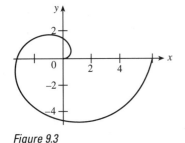

Figure 9.3

Archimedes gave a construction for drawing a tangent in his book 'Περι ελικων', which reads 'peri helikon' and means 'about spirals'.

A spiral was defined in terms of motion. Archimedes wrote:

> If a straight line of which one extremity remains fixed be made to revolve at a uniform rate in a plane until it returns to the position from which it started, and if, at the same time as the straight line revolves, a point move at a uniform rate along the straight line, starting from the fixed extremity, the point will describe a spiral in the plane.

Figure 9.3 shows such a spiral, called an Archimedes spiral, following these rules.

1 a   The straight line starts in the position of the *x*-axis, and rotates at 1 rad/s. The point starts at the origin and moves out along the line at a speed of 1 m/s. Show that you can write the Cartesian parametric equations of this Archimedes spiral in the form $x = t \cos t$, $y = t \sin t$.

b Find the polar equation of the same Archimedes spiral.

2 Suppose that the straight line in question **1** rotates at ω rad/s, and that the point moves along the line with speed v m/s. Find its Cartesian parametric and polar equations.

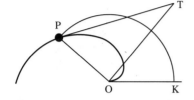

Figure 9.4

Here is the construction Archimedes gave for the tangent at the point P on an Archimedes spiral (see Figure 9.4).

● Draw an arc with centre O and radius OP to cut the initial line at K.
● Draw OT at right angles to the radius vector OP and equal in length to the circular arc PK.
● PT will be the tangent to the spiral at P.

3 a For the spiral in question **2**, let P be the point reached after t seconds. Find the coordinates of the point T according to Archimedes's construction.

b Find the gradient of the line PT and show that it has the same gradient as the tangent at P.

There is some evidence that Archimedes knew the parallelogram rule for adding vectors. The spiral is defined by two motions which you can treat as vectors: the velocity v of the point P along OP, and the angular velocity ω of the line OP around the point O, which gives rise to a velocity of $r\omega$ perpendicular to OP.

Figure 9.5 shows how these motions are combined vectorially. The resultant motion at P is along the tangent line at P.

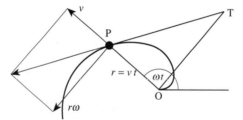

Figure 9.5

4 Use similar triangles to show that $OT = r\omega t$. Show that this is the length of the arc PK in Figure 9.4.

Some historians have argued that, since Archimedes was apparently relating motion and tangents, his method ought be described as differentiation. Others believe that the lack of generality does not justify this view. However, the relationship between tangent and motion was a fruitful concept when mathematicians in the 17th century began to work on the problem of finding tangents.

Tangents in the 17th century

Fermat, Descartes's contemporary, who was familiar with Archimedes's work on spirals, used the Greek definition of tangent in a more general method based on a way of finding maxima and minima for constructing tangents.

You will investigate Fermat's method for maxima and minima in Activity 9.3, moving on to tangents in Activity 9.4. However, Fermat's argument is difficult to follow; his reasoning is very unclear. The argument in Activity 9.3 is based on Fermat's writings, but it is unlikely to be the argument that Fermat actually used.

Activity 9.3   Fermat and maxima

1   Suppose that you have to divide a straight line of length b into two parts so that the product z of the two parts is a maximum. How should you do it?

Let x be one of the segments of the line b. Then $z = x(b - x)$.

Fermat decided to look at values of z close to x, so he replaced x by $x + e$, where e is small, to get a new value for the product. Call this new product z'.

a   Write down an expression for z'.

Fermat was trying to find the maximum value of the number z. He argued that if the maximum value happens at x, then at $x + e$ the value must be smaller than at x. Therefore $z - z'$ is always positive.

b   Find an expression for $z - z'$.

Fermat then goes on to use an argument similar to the following. If $z - z' > 0$ for all values of e, then $e^2 + e(2x - b) > 0$ for all values of e. This can only occur if the coefficient of e is zero. Therefore $x = \frac{1}{2}b$.

c   Explain why the coefficient of e has to be zero.

2   Fermat then gives a second, more complicated, example. In this example he still knows where the maximum value occurs, perhaps from numerical or graphical experimentation, but he is trying to prove his result. Again he makes a substitution.

Suppose that you are trying to find the maximum value of the expression $z = a^2 x - x^3$ as x varies, and suppose that it happens when $x = X$. Then $z = a^2 X - X^3$. For a value $x = X + e$, where e is small, the new value of $a^2 x - x^3$ must be smaller than the old one. Call this new value z'.

a   Write down an expression for z'.
b   Find and simplify an expression for $z - z'$.

The argument is now more complicated than for question **1**. As $z - z' > 0$ for all values of e which are small, the value of $e(3X^2 - a^2) + 3Xe^2 + e^3$ must be positive when e is small. When e is small, e^3 is very small compared with e^2 and can be ignored. Thus $z - z' \approx e(3X^2 - a^2) + 3Xe^2$. You can now use the argument of question **1**.

c   What does the argument of question **1** give for the value of X when z is a maximum? How do you resolve the difficulty which follows?
d   Check your answer to part **d** by differentiation.

To see how Fermat constructed tangents, work through Activity 9.4. Fermat's original argument was difficult to follow and virtually none of his contemporaries, including Descartes, understood it. Descartes used a similar method described in Activity 9.5, without realising that it was essentially the same as Fermat's.

Activity 9.4   Fermat and constructing tangents

This activity is about drawing the tangent TM to the curve $ax^2 = y^3$ at the point $M(x, y)$, shown in Figure 9.6. O is the origin.

Fermat's strategy was to find an expression for the length of the line NZ as Z moves along the curve, and then to use the fact that, for this curve, the length NZ, which is never negative, takes its minimum value of zero when Z coincides with the point M.

Call the distance $PT = s$, and let the distance PQ be e.

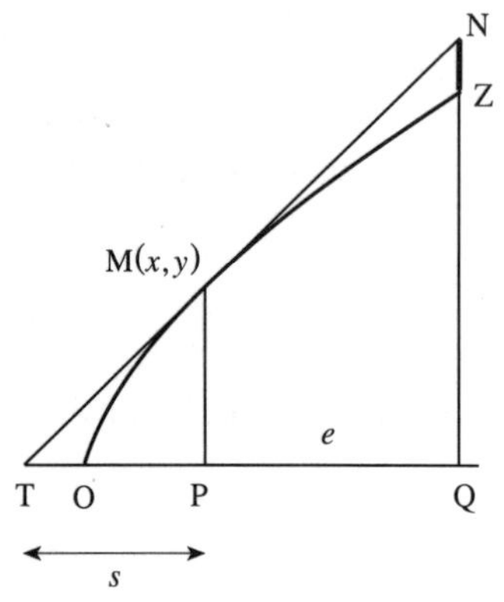

1 a Use similar triangles to find an expression for NQ in terms of y, s and e.
b Write down the x-coordinate of Q, and hence find the y-coordinate of Z.

2 You will find from your answers to question **1** that the expression for $ZN = (QN - QZ)$ is complicated. Find an expression for $QN^3 - QZ^3$ instead. Give your answer in terms of a, s, e and x. Show that the coefficient of e in your expression is $\dfrac{ax}{s}(3x - 2s)$.

3 Fermat now argues that this coefficient has to be zero. If it is not, he says, then the value of $QN^3 - QZ^3$ will become negative close to $e = 0$.
a Find the value of s in terms of x.
b Choose a value of a and draw an example of this curve on your calculator, and check Fermat's result.
c Use calculus to prove Fermat's result.

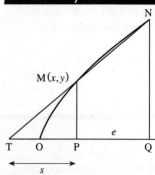

Figure 9.6

The connection between Activities 9.4 and 9.5 is that, in both cases, Fermat considered a difference, which always had to be positive. Descartes used a similar method, but directly on the construction of tangents. He also used a subtly different definition of tangent: a tangent cuts a curve in two coincident points. He thus used a slightly different diagram for his method, in which he also calculates the length TP.

Activity 9.5   Descartes's approach to tangents

1 a Write down the coordinates of N in terms of x, e and s.
b Write down the relationship between the coordinates of N, remembering that it now lies on the curve.
c Multiply out the equation you have written down, and write it showing the coefficients of e, e^2 and e^3.
d Divide through by e.

Descartes then argued that, when TM is a tangent, N and M coincide, so $e = 0$ is a solution of this equation.

2 Put $e = 0$ and find an expression for s.

Figure 9.7

Descartes published this method in 1637 in *La Géométrie*, which contains many of the ideas of differential calculus. You studied a different aspect of this book in the *Descartes* unit. Other mathematicians, among them Roberval, Barrow and Kepler, were also involved in similar work on calculus.

Quadrature

As you know from your work in *The Greeks* unit, quadrature originally meant constructing a square equal in area to a given shape. However, the work of Archimedes had changed the meaning, so that today it has come to mean finding a numerical value for area.

Archimedes came close to a modern method of integration when he found the area under a parabola. This is described in Activity 4.7 in *The Greeks* unit.

In 1635, long after Archimedes, Bonaventura Cavalieri, a professor at Bologna in Italy, published a treatise on a method of quadrature based on 'indivisibles'. He considered an area to be the sum of an infinite number of infinitely thin lines. Activity 9.6 shows how the method of indivisibles works.

Activity 9.6   *Using indivisibles*

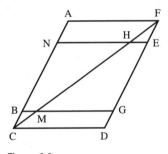

Figure 9.8

The abbreviation 'o.l.' means 'omnes lineae', or 'all lines'. In modern notation you would write $\sum a$ = area of ACDF.

1 Read through Cavalieri's argument, presented below. Make sure that you understand it. (It is not a valid argument, but do not spend too long looking for the error, which is subtle.)

The aim is to find an expression for the area of a triangle CDF. A congruent triangle AFC is drawn, making a parallelogram AFDC, as shown in Figure 9.8.

The lines BG and NE are drawn parallel to CD and AF. They meet the diagonal of the parallelogram at M and H. Notice that HE = BM and NH = MG.

Let HE = x, NH = y and AF = a. Then $x + y = a$.

If you were to sum all the possible parallel lines NE between AF and CD, then you could write, like Cavalieri

$$\text{o.l.}\ a = \text{area of ACDF}$$
$$= \text{o.l.}\ (x + y) = \text{o.l.}\ x + \text{o.l.}\ y$$

Since HE = BM, and for every such BM there is a unique and equal line HE, and similarly, since NH = MG and for every NH there is a unique and equal line MG,

$$\text{o.l.}\ x = \text{o.l.}\ y$$

Hence o.l. $x = \frac{1}{2}$ o.l. a.

But o.l. x is the area of triangle CFD, so area of CDF = $\frac{1}{2}$ area of ACDF.

Cavalieri was able to apply the method of indivisibles to a variety of curves and solids and derived the relationship now written as

$$\int_0^a x^n\, dx = \frac{a^{n+1}}{n+1}$$

where n is a positive integer. The abbreviation 'o.l.' was a form of integration symbol.

Cavalieri was aware that his method was built upon shaky ground; there was no firm mathematical proof that an area could be created by summing an infinite number of infinitely thin lines, but the method worked if he was careful, and he was sure that the problems would eventually be ironed out. Proof that he knew of the problems is revealed in a letter to his friend Torricelli, to whom he points out this paradox, which uses the same argument as in Activity 9.6.

Triangle ABC in Figure 9.9 is non-isosceles, and AD is an altitude. Clearly, the area of the triangle ABD does not equal the area of the triangle ADC.

However, PQ is parallel to BC, and PR is parallel to QS. Thus $PR = QS$.

But, for every PR there is a unique QS, so o.l. $PR = $ o.l. QS.

This implies that the area of triangle ABD = area of triangle ADC. By the method of indivisibles then, the two areas are both equal and at the same time unequal.

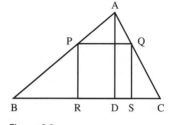

Figure 9.9

The work of Leibniz

Leibniz (1646–1716) was born in Leipzig. Although his family were well-to-do, Leibniz had no guaranteed income and had to work all his life. Between 1672 and 1675, when he was working in Paris as a diplomatic attaché for the Elector of Palatine, he visited London where he met some English mathematicians and was told of some ideas which Newton had used in developing the calculus. In 1677, after he had left Paris and was working in Brunswick, Leibniz published his own version of the calculus.

Subsequently Leibniz tried not to become embroiled in a dispute with Newton over who had first invented calculus. Although Newton spent considerable time and energy claiming precedence for the invention of the calculus, it is Leibniz's notation which we use today.

Leibniz initially used Cavalieri's notation, but in the period from 25th October to 11th November 1675 he invented the notation that we use today. He began much as Cavalieri did, and, over a period of a few days, improved the notation and also derived a number of rules just by playing with the symbols.

Activity 9.7   *Leibniz and area*

Leibniz started with a diagram similar to Figure 9.10. (Leibniz's diagram was actually rotated through 90° because it was customary at that time to draw curves with the origin at the top left.)

The vertical lines are all the same distance apart. Each vertical has length y. The difference in length between one vertical line and the next is called w. The area OCD is the sum of all the rectangles xw.

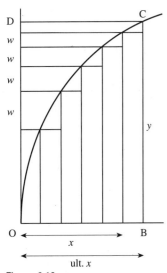

Figure 9.10

Leibniz wrote this as area OCD = omn. $\overline{xw}$.

> Leibniz wrote 'omn.' where Cavalieri had written 'o.l.' Leibniz also used the line over xw as a bracket.

1 Reproduce Leibniz's argument by writing down the expressions which fill the gaps.

area ODCB = ult. $x \times$ ult. y = ult. $x \times$ omn. ………… .

> Leibniz used 'ult. x' to mean the last term in the sequence of x's. 'Ult.' is short for the Latin 'ultimus', meaning last.

Now since each y is the sum of the w's so far, Leibniz can say

area OCB = omn. y = omn. ………… .

Putting these three results together, Leibniz has the formula

area OCD = area ODCD − area OCB

or: omn. ………… = ………… − ………….

2 Re-write the same argument using modern Σ notation.

Having got this formula, Leibniz worked on it.

Three days later he decided to drop the use of 'omn.' and began to use the symbol '$\int$', an elongated 's' standing for 'summa' meaning 'sum'. He worked out some rules for manipulating $\int$ such as $\int (x + y) = \int x + \int y$.

By 11th November 1675, Leibniz was considering quadrature to be the sum of rectangles of the form $y \times dx$ and was writing down recognisable formulae such as

$$\int y\,dy = \tfrac{1}{2} y^2$$

It was now clear to Leibniz that there was a connection between quadrature and finding tangents. You can see this by considering Activity 9.8. There, each value of y is the sum of all the w's so far. So, each w is the difference between two successive y's. Leibniz's method for finding tangents also used the differences between successive y values.

> It was not until later that Leibniz used this notation; at this point in the development he was writing $\int y$ for the area under the graph of y drawn against x.

Activity 9.8   Relating gradients and integration

Leibniz thought of dy, which he called a differential, as the difference between successive y-values, and dx as the difference between successive x-values.

Leibniz said that the differentials, dy and dx, are infinitely small, but can still be compared to each other, so the ratio $\dfrac{dy}{dx}$ is finite.

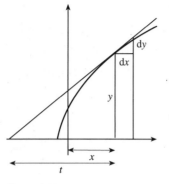

Figure 9.11

1 Write down a relation between the ratio $\dfrac{dy}{dx}$ and the ratio of y to the length t in Figure 9.11. Now think about the 'differential' of the area under the curve: that is, the difference between two successive values of the area.

2 Draw a diagram like Figure 9.11 to illustrate the difference between successive areas.

3 If the differential of the areas is written $d\displaystyle\int y\,dx$, to what is it equal?

Leibniz claimed that the result of question **3** showed that the operations 'd' and '$\displaystyle\int$' are inverse.

Leibniz now derived the familiar rules of differentiation using two observations.

- As differentials are infinitely small, they can be ignored compared with ordinary finite quantities, so $x + dx = x$.
- Products of differentials can be ignored, in comparison with the differentials themselves, so $a \times dx + dx \times dy = a \times dx$.

So, for example, Leibniz wrote

$$d(uv) = (u + du)(v + dv) - uv$$
$$= u \times dv + v \times du + du \times dv$$
$$= u \times dv + v \times du$$

since, according to him, the product $du \times dv$ is negligible.

> The work of Leibniz, while giving correct results and a useful notation, has been regarded as logically suspect, because it was based on the intuitive idea of differentials. In the 1870s, a German mathematician, Weierstrass, gave a rigorous treatment of calculus which did not involve differentials. However, in 1960, an American, Abraham Robinson, gave a precise treatment of calculus by using differentials.

Newton's approach

Isaac Newton (1642–1727) went to Grantham Grammar School and then to Trinity College, Cambridge. His teacher at Cambridge was Isaac Barrow, the first Lucasian Professor of Mathematics. He resigned in 1669 so that Newton could take over the post! It was Barrow's lectures on geometry and in particular on methods he had devised for finding areas and tangents that provided the foundation for Newton's inventions of integral and differential calculus. His first paper on fluxions was written in 1666 and he extended the work in a paper called *De methodis fluxionum* written in 1670–71. However, Newton was a secretive person, and kept his discoveries to himself. By not publishing his work, he left the way open for others to dispute whether or not he had priority over the invention of the calculus.

The peak of Newton's mathematical career was the publication in 1687 of a book called *Philosophiae Naturalis Principia Mathematica*, detailing all of his important mathematical and scientific discoveries. In 1693, recovering from a serious illness, Newton discovered that his calculus was being used in Europe, and that its invention was being attributed to Leibniz. In 1699 he was made Master of the Royal Mint and subsequently spent considerable energy trying to prove that Leibniz was a plagiarist. He continued the attack even after Leibniz had died!

Newton's approach to calculus was based on a view of curves different from Leibniz's idea, although the end result was in practice the same. Leibniz considered a curve to be a series of infinitely small steps, as we have seen. Newton, writing in *De quadratura,* says:

> I don't here consider Mathematical Quantities as composed of Parts extreamly small, but as generated by a continual motion. Lines are described, and by describing are generated, not by any apposition of Parts, but by a continual motion of Points. Surfaces are generated by the motion of Lines, Solids by the motion of surfaces, Angles by the Rotation of their Legs, Time by a continual flux, and so in the rest. These Geneses are founded upon Nature, and are every Day seen in the motion of Bodies.

Newton's work is based on two main ideas:
- fluents, which are variables that increase or decrease with time
- fluxions, the speeds of the fluents.

In Leibniz's notation, if x is a fluent, then the fluxion is $\dfrac{\mathrm{d}x}{\mathrm{d}t}$. Originally, Newton used a different letter to represent the fluxion of a fluent, but in 1691 he invented what he called 'pricked letters', even though he was aware of Leibniz's notation.

Newton then wrote $\dot{x}$, pronounced 'x dot', to mean the fluxion of a fluent x.

Here is an example of Newton's work which illustrates one of the fundamental problems associated with calculus in its infancy: the idea of a limit.

> Let the Quantity x flow uniformly, and let the Fluxion of x^n be to be found. In the same time that the Quantity x by flowing becomes $x + o$, the quantity x^n will become $(x + o)^n$, that is, by the Method of Infinite Series
>
> $$x^n + nox^{n-1} + \frac{n(n-1)}{2}oox^{n-2} +, \& c$$ and the Augments o and
>
> $$nox^{n-1} + \frac{n(n-1)}{2}oox^{n-2} +, \& c$$ are to one another as 1 and
>
> $$nx^{n-1} + \frac{n(n-1)}{2}ox^{n-2} +, \& c.$$ Now let those Augments vanish and their ultimate ratio will be the Ratio of 1 to nx^{n-1}; and therefore the Fluxion of the Quantity x is to the Fluxion of the Quantity x^n as 1 to nx^{n-1}.

The Method of Infinite Series is Newton's term for what you know as the binomial series; Newton invented it in the winter of 1664–5.

Fluents and fluxions have already been defined; but you also need to know that an 'augment' is a small increment, or increase, and the 'ultimate ratio' is the ratio of the quantities as the augments vanish.

Activity 9.9   Newton and differentiation

1   Write Newton's argument above in modern notation.

The problem with Newton's basis for calculus, although it worked, was that it was not clear whether you could actually let the augments vanish. If they did vanish, how could they have a ratio? Leibniz avoided this problem to some extent by using his infinitely small increments dx and dy, which never became zero, but which were so small that they could be ignored with respect to the quantities x and y.

The dispute over Newton's vanishing augments, he called them 'evanescent augments', raged on for many years after his death. It was only resolved in the 19th century when Cauchy developed the idea of a limit.

Reflecting on Chapter 9

What you should know

- that calculus provides a general method for solving problems that previously required diverse special techniques
- some early methods for finding tangents
- examples from the work of each of Archimedes, Fermat and Descartes
- some earlier methods for finding areas enclosed by curves
- an example from the work of each of Archimedes and Cavalieri
- the distinctive approach taken by Newton, his notation, and some examples
- the logical difficulties inherent in his approach
- the distinctive approach taken by Leibniz, his notation, and some examples
- the logical difficulties inherent in his approach.

Preparing for your next review

- Reflect on the 'What you should know' list for this chapter. Be ready for a discussion on any of the points.
- Answer the following check question.

1 Fermat had determined the area under any curve of the form $y = x^k$ and, in the 1640s, had been able to construct the tangent to such a curve. Use any sources at your disposal to discover why it is that calculus is not considered 'invented' until the time of Newton and Leibniz.

Practice exercises for this chapter are on page 156.

6

Searching for the abstract

Introduction

Chapter 10
'Two minuses
make a plus'

Chapter 11
Towards a
rigorous approach

In this unit you will see an example of how, over time, a piece of mathematics changes. At first the mathematics gets created with little or no justification or proof that the mathematics is 'correct'. With time, or because its use changes, so the mathematics becomes more formally established. It gets 'proved'. This may remind you of the process of investigating and proving which Book 4, Chapter 16, encouraged you to adopt. You make conjectures and work on them, often using examples, pictures and diagrams. Only later do you get round to proving your conjectures. This process of proof is sometimes called formalising. When mathematics is formalised, it can also become extended and generalised.

In Chapter 10 you will follow the development of negative numbers and rules for combining them from their early use in China in the 3rd century BC through to 18th century Europe. In Chapter 11 you will discover how negative numbers became 'respectable' and how the rules for combining all numbers, not just negative ones, were put on a firm footing. With this status, number systems can be further developed into more elaborate algebraic systems, which share some properties with number systems. You will follow the ideas of Wessel and Clifford, who both proposed extensions to the number system.

This unit is designed to take about 10 hours of your learning time. About half of this time will be outside the classroom.

It is more important that you understand the general message of these two chapters, rather than becoming too involved in all the details of the activities.

Work through the chapters in sequence.

There are summaries and further practice exercises in Chapter 12.

Mathematical knowledge assumed

- you will need to know a little about vectors in both two and three dimensions
- it is also helpful, but not essential, to have a little knowledge of complex numbers, but any information that you need is explained in the unit.

10   'Two minuses make a plus'

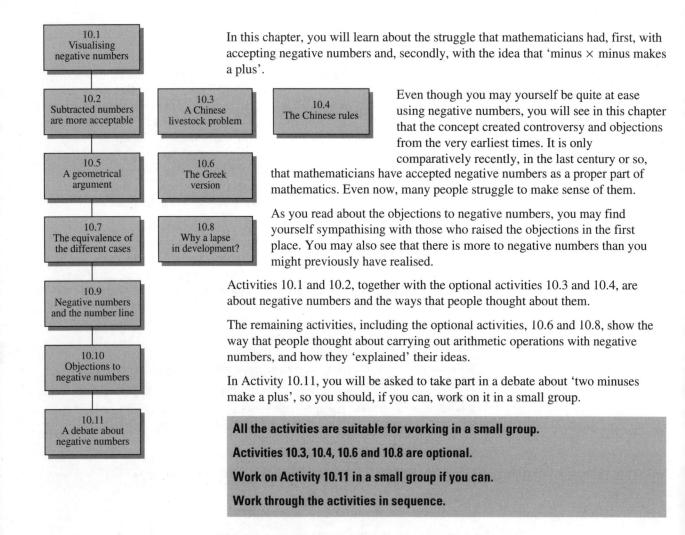

10.1
Visualising
negative numbers

10.2
Subtracted numbers
are more acceptable

10.3
A Chinese
livestock problem

10.4
The Chinese rules

10.5
A geometrical
argument

10.6
The Greek
version

10.7
The equivalence of
the different cases

10.8
Why a lapse
in development?

10.9
Negative numbers
and the number line

10.10
Objections to
negative numbers

10.11
A debate about
negative numbers

In this chapter, you will learn about the struggle that mathematicians had, first, with accepting negative numbers and, secondly, with the idea that 'minus × minus makes a plus'.

Even though you may yourself be quite at ease using negative numbers, you will see in this chapter that the concept created controversy and objections from the very earliest times. It is only comparatively recently, in the last century or so, that mathematicians have accepted negative numbers as a proper part of mathematics. Even now, many people struggle to make sense of them.

As you read about the objections to negative numbers, you may find yourself sympathising with those who raised the objections in the first place. You may also see that there is more to negative numbers than you might previously have realised.

Activities 10.1 and 10.2, together with the optional activities 10.3 and 10.4, are about negative numbers and the ways that people thought about them.

The remaining activities, including the optional activities, 10.6 and 10.8, show the way that people thought about carrying out arithmetic operations with negative numbers, and how they 'explained' their ideas.

In Activity 10.11, you will be asked to take part in a debate about 'two minuses make a plus', so you should, if you can, work on it in a small group.

All the activities are suitable for working in a small group.

Activities 10.3, 10.4, 10.6 and 10.8 are optional.

Work on Activity 10.11 in a small group if you can.

Work through the activities in sequence.

Negative numbers

Activity 10.1   Visualising negative numbers

1   You already have different ways of thinking about and visualising negative

numbers. The same is true of people across the world and across the centuries. But why did people consider the possibility of negative numbers in the first place? What practical use do you think negative numbers have?

2   You can visualise numbers by considering them as points on a number line. The integers are equally spaced and increase from left to right. You can model an operation combining numbers as a movement along the line; the number which is the result of the operation corresponds to the point where the movement ends.

a   How can you interpret the operations of addition and subtraction as movements along the number line?

 b   What about multiplication? Does your picture of multiplication include the multiplication of a positive number by a negative number and a negative by a positive? Does it extend further to a negative by a negative? If not, why?

Before you read more about the changing views about negative numbers through history, consult the answers for this activity to find out more about the use to which negative numbers were put and how the number line can be used to model operations on numbers.

Negative numbers have a chequered history. Although they were found as solutions of equations as early as the 3rd century BC, they were usually rejected because they were not considered to correspond to proper solutions of practical problems. Negative numbers were called 'absurd' by Diophantus in the 3rd century and by Michael Stifel, a German algebraist, in the first half of the 16th century, and 'fictitious' by Geronimo Cardano, again in the 16th century. Descartes called negative solutions of equations 'racines fausses', literally 'false roots'. The Chinese used the word 'fu' in front of the word for number when describing negatives. 'Fu' is like the English prefix 'un' or 'dis'. This re-inforces the general impression that, for a long time, negative numbers were not thought of as proper numbers.

Activity 10.2   *Subtracted numbers are more acceptable*

Historically, the concept of a 'subtracted number' was more common than that of a negative number. A subtracted number is one such as $a - b$, where $a > b$.

1   People found subtracted numbers acceptable even when they rejected negative numbers out of hand. Why do you think that this was so?

Subtracted numbers, and operations on them, can be traced back to civilisations that you studied at the beginning of this book.

2   Find some examples of the use of subtracted numbers in *The Babylonians* and *The Greeks* units. Can you propose some other uses to which they could be put?

Subtracted numbers appeared in equations which were considered, and solved, by the Babylonians, Chinese and Indians.

If you look at such equations today, you might interpret them differently by believing that they contain negative terms, that is terms with negative coefficients.

For example, if an equation contains the expression $a - 2b$, you could consider it as $a + (-2)b$, where -2, the coefficient of b, is a negative number.

However the Babylonian, Chinese and Indian mathematicians viewed $a - 2b$ as $2b$ being subtracted from a. If their equations contained $a - 2b$, they were only prepared to consider those solutions in which $a - 2b$ was positive.

Activity 10.3   A Chinese livestock problem

Examples of equations with subtracted numbers are to be found in a Chinese text dating back to about the 3rd century BC, called the *Chiu-chang suan-shu*, or the 'Nine chapters on the mathematical arts'.

One example is from a problem involving payment for livestock. Selling 2 sheep and 5 pigs while buying 13 cows leaves a debt of 580 pieces of money; selling 3 cows and 3 pigs while buying 9 sheep leaves no money at all; and buying 5 cows but selling 6 sheep and 8 pigs leaves 290 pieces.

For solving problems that involved what you would call simultaneous equations, the Chinese used algorithms carried out on a counting board with coloured rods: red for added numbers and black for subtracted ones. On the counting board the problem would be set out as

-5	3	-13	cows
6	-9	2	sheep
8	3	5	pigs
290		-580	pieces

 1 Set out the Chinese livestock problem as a system of simultaneous equations and solve it using the method you developed in Chapter 5, in the *Equations and inequalities* unit in Book 3.

Ways of combining negative numbers

The *Chiu-chang suan-shu* gives rules for adding and subtracting positive and negative numbers. However, the rules are somewhat indirect because signs for the numbers were not used.

Activity 10.4   The Chinese rules

This activity is optional.

Here is an extract from the *Chiu-chang suan-shu*.

> When the equally signed quantities are to be subtracted and the different signed are to be added (in their absolute values), if a positive quantity has no opponent, make it negative; and if a negative has no opponent make it positive. When the different signed quantities are to be subtracted and the

equally signed are to be added (in absolute values), if a positive quantity has no opponent, make it positive; and if a negative has no opponent, make it negative.

1 Translate the rules in the extract for the treatment of positive and negative numbers into rules which your fellow students would understand. Use whatever means you like to explain these rules – words, pictures, tables, and so on – but keep the idea of the counting board and the way that the Chinese used it.

Although the Chinese probably provided the earliest known example of the use of negative numbers, and as early as the 3rd century BC had rules for adding and subtracting them, they were not the first to generate the rules for multiplying and dividing negative numbers. In China, these rules did not appear until 1299, in a work called the *Suan-hsiao chi-meng*.

The first known setting out of the rules for multiplying and dividing negative numbers was in the 7th century AD. The Indian Brahmagupta wrote down these rules which are instantly recognisable.

Positive divided by positive, or negative by negative is affirmative … . Positive divided by negative is negative. Negative divided by affirmative is negative.

Unlike those before him, Brahmagupta was prepared to consider negative numbers as 'proper', valid solutions of equations.

Activity 10.5   A geometrical argument

> You first learned about al-Khwarizmi in *The Arabs* unit in Chapter 5.

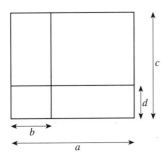

Figure 10.1

The rules presented by Brahmagupta had already been considered by the Greeks some four centuries earlier. Theirs was a geometrical argument, so differences in distances and areas were used and were represented as subtracted numbers. The geometrical setting is presented here in a slightly different form from the way it was set out originally in *Book II* of Euclid's *Elements*. It is as argued by al-Khwarizmi in the 9th century, and it should look familiar to you from your previous work on Chapter 2 of the *Investigating and proving* unit in Book 4.

1 Explain carefully in words how Figure 10.1 enables you to conjecture the algebraic formula

$$(a-b)(c-d) = ac - bc - ad + bd.$$

In the algebraic expression that you obtained in Activity 10.5, the values of a, b, c and d are all positive and are such that $a > b$ and $c > d$, so you used positive numbers throughout. There are no negative numbers present, only subtracted numbers. Even if you subtracted some of these positive numbers, you made no use of negative numbers standing on their own. Also, although you can very easily use this expression to get a rule for multiplying negative numbers, this was not done at the time, since negative numbers were assumed not to exist.

Activity 10.6 The Greek version

This activity is optional.

Proposition 7 from *Book II* of Euclid's *Elements* (in English translation, 1726) describes a result closely related to that of al-Khwarizmi above.

PROPOSITION VII.
A PROBLEM.

If a Line be divided, the Square of the whole Line with that of one of its Parts, is equal to two Rectangles contain'd under the whole Line, and that first Part, together with the Square of the other Part.

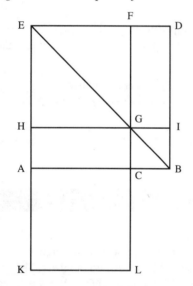

Let the Line AB be divided anywhere in C; the Square AD of the Line AB, with the Square AL, will be equal to two Rectangles contain'd under AB and AC, with the Square of CB. Make the Square of AB, and having drawn the Diagonal EB, and the Lines CF and HGI; prolong EA so far, as that AK may be equal to AC; so AL will be the Square of AC, and HK will be equal to AB; for HA is equal to GC, and GC, is equal to CB, because CI is the Square of CB *by the Coroll. of the 4.)*

Demonstration.

'Tis evident, that the Squares AD and AL are equal to the Rectangles HL and HD, and the Square CI. Now the Rectangle HL is contain'd under HK equal to AB, and KL equal to AC. In like manner the Rectangle HD is contain'd under HI equal to AB, and HE equal to AC. Therefore the Squares of AB and AC are equal to two Rectangles contain'd under AB and AC, and the Square of CB.

In Numbers.

Suppose the Line AB to consist of 9 Parts, AC of 4, and CB of 5. The Square of AB 9 is 81, and that of AC 4 is 16; which 81 and 16 added together make 97. Now one Rectangle under AB and AC, or four times 9, make 36, which taken twice, is 72; and the Square of CB 5 is 25; which 72 and 25 added together make also 97.

 1 a In your own words, explain the result of this proposition and also the argument leading to it and write down the result of the proposition as an equation.

b What connects this result and that of al-Khwarizmi in Activity 10.5?

The geometrical argument given by Euclid in Activity 10.6 with the later variant by al-Khwarizmi (Activity 10.5) re-appeared with slight variations time and again over the centuries. For example, it appeared in the work of the French civil servant, François Viète, who, you will remember from the *Descartes* unit, did mathematics in his spare time. He reproduced al-Khwarizmi's result, but in a very different form, using words and making substantial use of symbols to represent the quantities, rather than using a geometric argument.

This is an interesting case of how a result is generated first by considering a diagram, where the result is apparent from the picture, and then, sometime later, is turned into a symbolic formula. Viète's symbolic explanation of al-Khwarizmi's geometric result was probably the first algebraic formulation of a geometric argument. It is algebraic because it uses symbols which are not necessarily assumed to represent distances.

Activity 10.7   The equivalence of the different cases

One dictionary gives the meaning of extrapolate as 'to infer, conjecture, from what is known'.

 1 How would you use al-Khwarizmi's equation, which you obtained in Activity 10.5, to extrapolate rules for multiplying two negative numbers together?

 2 Even though al-Khwarizmi did not express it in quite the same way, his equation is equivalent to Brahmagupta's rules as set out on page 127. This is because al-Khwarizmi's equation enables you to derive the rules. Explain how you could obtain the rules from al-Khwarizmi's equation by extrapolation.

As you can see from Activity 10.7, it is a simple extension of al-Khwarizmi's equation to deduce the result of multiplying two negative numbers together. However this was not done by any of the mathematicians mentioned so far, because it required them to admit the idea of a negative number.

It appears to have been the Flemish mathematician, Albert Girard (1590–1633), who first openly acknowledged the validity of negative roots of equations and thus opened up the debate about the existence of negative numbers. He interpreted negative numbers as having an opposite direction to positive ones – he used the words 'retrogression' and 'advance' for negative and positive numbers respectively – which might remind you of where this chapter began with the number line.

Activity 10.8   Why a lapse in development?

This activity is optional.

From Girard's recognition of negative numbers in 1629 there is a lag in time to the mid-18th century before the next significant developments regarding rules for combining negative numbers.

 1 From your knowledge of what was going on in the mathematics community during that time, can you propose any reason which might account for this lapse?

Although the rules for combining negative numbers were not being developed between Girard's time and the mid-18th century, negative numbers themselves were not totally neglected, since further progress was being made on ordering of numbers along the number line.

Searching for the abstract

John Wallis (1616–1703), a contemporary of Newton, had a healthy scepticism about the rules for dividing negative numbers. He made considerable progress in interpreting numbers as points on the number line.

Activity 10.9 Negative numbers and the number line

In 1673, Wallis wrote the following in his book entitled *Algebra*.

But it is also Impossible, that any Quantity (though not a Supposed Square) can be *Negative*. Since that it is not possible that any *Magnitude* can be *Less than Nothing*, or any *Number Fewer than None*.

Yet is not that Supposition (of Negative Quantities,) either Unuseful or Absurd; when rightly understood. And though, as to the bare Algebraick Notation, it import a Quantity less than nothing: Yet, when it comes to a Physical Application, it denotes as Real a Quantity as if the Sign were + ; but to be interpreted in a contrary sense.

As for instance: Supposing a man to have advanced or moved forward, (from A to B,) 5 Yards; and then to retreat (from B to C) 2 Yards: If it be asked, how much he had Advanced (upon the whole march) when at C? or how many Yards he is now Forwarder than when he was at A? I find (by Subducting 2 from 5,) that he is Advanced 3 Yards. (Because + 5 – 2 = + 3.)

But, if having Advanced 5 Yards to B, he thence Retreat 8 Yards to D; and it then be asked, How much he is Advanced when at D, or how much Forwarder than

when he was at A: I say – 3 Yards. (Because + 5 – 8 = – 3.) That is to say, he is advanced 3 Yards less than nothing.

Which in propriety of Speech, cannot be, (since there cannot be less than nothing.) And therefore as to the Line AB *Forward*, the case is Impossible.

But if, (contrary to the Supposition,) the Line from A, be continued *Backward*, we shall find D, 3 Yards *Behind* A. (Which was presumed to be *Before* it.)

And thus to say, he is *Advanced* – 3 Yards; is but what we should say (in ordinary form of Speech), he is *Retreated* 3 Yards; or he wants 3 Yards of being so Forward as he was at A.

Which doth not only answer Negatively to the Question asked. That he is not (as was supposed,) Advanced at all: But tells moreover, he is so far from being advanced, (as was supposed) that he is Retreated 3 Yards; or that he is at D, more Backward by 3 Yards, than he was at A.

And consequently – 3, doth as truly design the Point D; as + 3 designed the Point C. Not Forward, as was supposed; but Backward, from A.

1 What do you believe is noteworthy about this passage?

In this passage, Wallis uses the + and – signs in front of the numbers to indicate a particular direction, not to convey the idea of addition and subtraction. This is the first example in this chapter of signs representing anything other than the usual operations of combining numbers. More examples of this directional use will follow.

Wallis was not the only person to have concerns about the rules for combining negative numbers. At exactly the same time and also using the number line, the French mathematician Antoine Arnauld (1612–1694) presented arguments which he believed showed the rules about negative numbers to be contradictory.

Activity 10.10   *Objections to negative numbers*

In the following extract, Arnauld raises strong objections to the use of negative numbers.

> … je ne comprens pas que le quarré de –5 puisee être la mème chose que le quarré de +5, et que l'un et l'autre soit +25, Je ne sçai de plus comment ajuster cela au fondement de la multiplication, qui est que l'unité doit être à l'une des grandeurs que l'on multiplie, comme l'autre est au produit. Ce qui est également vrai dans les entiers et dans les fractions. Car 1 est à 3, comme 4 est à 12. Et 1 est à $\frac{1}{3}$ comme $\frac{1}{4}$ est à $\frac{1}{12}$. Mais je ne puis ajuster cela aux multiplications de deux moins. Car dira-t-on que +1 est à –4, comme –5 est à +20? Je ne le vois pas. Car +1 est plus que –4. Et au contraire –5 est moins que +20. Au lieu que dans toutes les autres proportions, si le premier terme est plus grand que le second, le troisiéme doit être plus grand que le quatriéme.

1 a   Translate the extract from Arnauld's letter into English. If you believe that your French is not up to it, then ask the help of a friend – but not before having tried to get an impression for yourself of what Arnauld claims. (You should only refer to the hint as a last resort!)

b   Which properties of numbers is Arnauld using in his argument?

c   From the point of view of modern-day mathematics, what is the fallacy in his argument? What part of your A-level course can you use to argue against Arnauld?

It took until the middle of the 18th century for further developments to take place; then they accelerated. Leonard Euler (1707–1783) and Nicholas Saunderson (1682–1739) were partly responsible for this. They produced some of the first algebra textbooks, which dwelt at some length on the rules for multiplying negative numbers. This emphasis was quite natural, because negative numbers and their rules of combination had not been described in texts until that time. Both texts were intended for teaching and were considered well designed for this purpose, and many editions of each of them were published. It has been said that the textbooks were easily understandable partly because of their authors' blindness; Euler, for example, had to dictate his text and make it intelligible to a relatively unskilled person.

> Interestingly Euler and Saunderson were both blind, Euler only for the last 17 years of his life, but Saunderson from the age of 1, when he suffered from smallpox.

Activity 10.11   *A debate about negative numbers*

The following is an extract from Euler's *Elements of Algebra*, which first appeared in 1774. Many editions were published centuries later. This one is from the 1828 edition.

31. Hitherto we have considered only positive numbers; and there can be no doubt, but that the products which we have seen arise are positive also: viz. $+a$ by $+b$ must necessarily give $+ab$. But we must separately examine what the multiplication of $+a$ by $-b$, and of $-a$ by $-b$, will produce.

32. Let us begin by multiplying $-a$ by 3 or $+3$. Now, since $-a$ may be considered as a debt, it is evident that if we take that debt three times, it must thus become three times greater, and consequently the required product is $-3a$. So if we multiply $-a$ by $+b$, we shall obtain $-ba$, or, which is the same thing, $-ab$. Hence we conclude, that if a positive quantity be multiplied by a negative quantity, the product will be negative; and it may be laid down as a rule, that $+$ by $+$ makes $+$ or *plus*; and that, on the contrary, $+$ by $-$, or $-$ by $+$, gives $-$, or *minus*.

33. It remains to resolve the case in which $-$ is multiplied by $-$; or, for example, $-a$ by $-b$. It is evident, at first sight, with regard to the letters, that the product will be ab; but it is doubtful whether the sign $+$, or the sign $-$, is to be placed before it; all we know is, that it must be one or the other of these signs. Now, I say that it cannot be the sign $-$: for $-a$ by $+b$ gives $-ab$, and $-a$ by $-b$ cannot produce the same result as $-a$ by $+b$; but must produce a contrary result, that is to say $+ab$; consequently, we have the following rule: $-$ multiplied by $-$ produces $+$, that is the same as $+$ multiplied by $+$.

The next extract is from Saunderson's *Elements of Algebra*, which ran to five editions between 1740 and 1792.

Of the multiplication of algebraic quantities.

And first, how to find the sign of the product in multiplication, from those of the multiplicator and multiplicand given.

5. Before we can proceed to the multiplication of algebraic quantities, we are to take notice, that if the signs of the multiplicator and multiplicand be both alike, that is, both affirmative, or both negative, the product will be affirmative, otherwise it will be negative: thus $+4$ multiplied into $+3$, or -4 by -3 produces in either case $+12$: but -4 multiplied into $+3$, or $+4$ into -3 produces in either case -12.

If the reader expects a demonstration of this rule, he must first be advertised of two things: *first*, that numbers are said to be in arithmetic progression, when they increase or decrease with equal differences, as 0, 2, 4, 6; or 6, 4, 2, 0; also as 3, 0, -3; 4, 0, -4; 12, 0, -12; or -12, 0, 12: whence it follows, that three terms are the fewest that can form an arithmetical progression; and that of these, if the two first terms be known, the third will easily be had: thus if the two first terms be 4 and 2, the next will be 0; if the two first be 12 and 0, the next will be -12; if the two first be -12 and 0, the next will be $+12$, &c.

2dly, If a set of numbers in arithmetical progression, as 3, 2 and 1, be successively multiplied into one common multiplicator, as 4, or if a single number, as 4, be successively multiplied into a set of numbers in arithmetic progression, as 3, 2 and 1, the products 12, 8 and 4, in either case, will be in arithmetical progression.

This being allowed, (which is in a manner self-evident,) the rule to be demonstrated resolves itself into four cases:

1*st*, That $+4$ multiplied into $+3$ produces $+12$.
2*dly*, That -4 multiplied into $+3$ produces -12.
3*dly*, That $+4$ multiplied into -3 produces -12.
And *lastly*, that -4 multiplied into -3 produces $+12$. These cases are generally expressed in short thus: first $+$ into $+$ gives $+$; secondly $-$ into $+$ gives $-$; thirdly $+$ into $-$ gives $-$; fourthly $-$ into $-$ gives $+$.

Case 1st. That $+4$ multiplied into $+3$ produces $+12$, is self-evident, and needs no demonstration; or if it wanted one, it might receive it from the first paragraph of the 3d article: for to multiply $+4$ by $+3$ is the same thing as to add $4+4+4$ into one sum; but $4+4+4$ added into one sum give 12, therefore $+4$ multiplied into $+3$, gives $+12$.

Case 2d. And from the second paragraph of the 3d art. it might in like manner be demonstrated, that -4 multiplied into $+3$ produces -12: but I shall here demonstrate it another way, thus: multiply the terms of this arithmetical progression 4, 0, -4, into $+3$, and the products will be in arithmetical progression, as above; but the first two products are 12 and 0; therefore the third will be −12; therefore −4 multiplied into +3, produces −12.

Case 3*d*. To prove that +4 multiplied into −3 produces −12; multiply +4 into +3, 0 and −3 successively, and the products will be in arithmetical progression; but the first two are 12 and 0, therefore the third will be −12; therefore +4 multiplied into −3, produces −12.

Case 4*th*. Lastly, to demonstrate, that −4 multiplied into −3 produces +12, multiply −4 into 3, 0 and −3 successively, and the products will be in arithmetical progression; but the two first products are −12 and 0, by the second case; therefore the third product will be +12; therefore −4 multiplied into −3. produces +12.

Cas. 2nd	+4,	0,	−4
	+3,	+3,	+3
	+12,	0,	−12.

Cas. 3d	+4,	+4,	+4
	+3,	0,	−3
	+12,	0,	−12.

Cas. 4th	−4,	−4,	−4
	3,	0,	−3
	−12,	0,	+12.

These 4 cases may be also more briefly demonstrated thus: +4 multiplied into +3 produces +12; therefore −4 into +3, or +4 into −3 ought to produce something contrary to +12, that is, −12: but if −4 multiplied into +3 produces −12, then −4 multiplied into −3 ought to produce something contrary to −12, that is, +12; so that this last case, so very formidable to young beginners, appears at last to amount to no more than a common principle in Grammar, to wit, that two negatives make an affirmative; which is undoubtedly true in Grammar, though perhaps it may not always be observed in languages.

If possible, work on this activity in groups of three to six. One or two students should designate themselves as representing Euler, another student Saunderson, and another student Arnauld, as introduced above. Then work through the activities below. If you are working on your own, alternative activities are suggested.

1   a   Everyone should read through the extracts from Euler's and Saunderson's books. Then, together, draw up a list of assumptions made by each author to enable him to derive the rules relating to the multiplication of negative numbers.

b   If you are representing either Euler or Saunderson, imagine yourself in his place and prepare to defend his position against the objections raised by Arnauld. If you are representing Arnauld, prepare to argue your case against either Euler or Saunderson. Stage a debate in front of your fellow students and ask them to judge who makes the most convincing case.

If you are working on your own, but within a group of other mathematics students, prepare an oral presentation summarising Arnauld's views and either Euler's or Saunderson's. Make the presentation to the other students. As part of your preparation, make clear which 'side' you would be on and why.

Otherwise, write an essay summarising Arnauld's views and either Euler's or Saunderson's. Give your own opinions on the validity of the respective arguments.

If you were working in groups on Activity 10.11, you may have reached the position at the end of the debate that your fellow students were unable to choose which pair made the most convincing argument. As Euler or Saunderson, you might have found it difficult to convince Arnauld that his objections to negative numbers were incorrect. Likewise, as Arnauld, you might have found it hard to establish that your objections really do correspond to a flaw in Euler's or Saunderson's arguments.

If you were working on your own on Activity 10.11 you might have come to a conclusion that amounts to no more than a 'difference of opinion'. Both Euler and Saunderson were prepared to accept rules of combination of negative numbers which Arnauld was not prepared to do.

With the benefit of hindsight, you are now able to look back on the Euler, Saunderson and Arnauld dispute and comment that they were using different rules of arithmetic. Arnauld's work was less extensive than that of the other two since it did not include multiplication of negative numbers. Each of their rules was valid; they were different, that is all. You are going to investigate in the next chapter how the rules for combining numbers were established and further developed.

Before you move on to the next chapter you might be interested to read the following extract by the French novelist Stendhal (1783–1842) on the acceptance, or not, of rules of arithmetic. Do you have any sympathy with Stendhal's view?

> My enthusiasm for mathematics may have had as its principal basis my loathing for hypocrisy, which for me meant my aunt Séraphie, Mme Vignon and their priests.
>
> In my view, hypocrisy was impossible in mathematics and, in my youthful simplicity, I thought it must be so in all the sciences to which, as I had been told, they were applied. What a shock for me to discover that nobody could explain to me how it happened that: minus multiplied by minus equals plus ! (This is one of the fundamental bases of the science known as *algebra*.)
>
> Not only did people not explain this difficulty to me (and it is surely explainable, since it leads to truth) but, what was much worse, they explained it on grounds which were evidently far from clear to themselves.
>
> M. Chabert, when I pressed him, grew confused, repeating his lesson, that very lesson to which I had raised objections, and everybody seemed to tell me: "But it's the custom; everybody accepts this explanation. Why, Euler and Lagrange, who presumably were as good as you are, accepted it!" [...]
>
> It was a long time before I convinced myself that my objection that $- \times - = +$ simply couldn't enter M. Chabert's head, that M. Dupuy would never reply to it save by a haughty smile, and that the *brilliant ones* to whom I put my questions would always make fun of me.
>
> I was reduced to what I still say to myself today: It must be true that $- \times -$ equals $+$, since evidently, by constantly using this rule in one's calculations one obtains results *whose truth cannot be doubted.*
>
> My great worry was this: Let RP be the line separating the positive from the negative, all that is above it being positive, all that is below negative;
>
>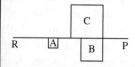
>
> how, taking the square B as many times as there are units in square A, can one make it change over to the side of square C?

> And, to use an awkward comparison which M. Chabert's pronounced Grenoblois drawl made even more clumsy, let us suppose that the negative quantities are a man's debts; how, by multiplying a debt of 10,000 francs by 500 francs, can this man have, or hope to have, a fortune of 5,000,000 francs?
>
> Are M. Dupuy and M. Chabert hypocrits like the priests who come to say Mass at my grandfather's, and can my beloved mathematics be a fraud? Oh, how eagerly I would have listened then to one word about logic, or the art of *finding out the truth*!

Reflecting on Chapter 10

What you should know

- negative numbers were initially rejected by the Chinese, Greeks and Babylonians, among others, as not being practical
- subtracted numbers on the other hand were acceptable to these people and, likewise, subtraction as an operation was considered acceptable
- rules for addition and subtraction of positive and negative numbers appeared as early as the 3rd century BC in China
- rules for multiplying subtracted numbers appeared later in the 3rd century AD in geometric form and in algebraic form in the 9th century (al-Khwarizmi)
- rules for multiplying and dividing negative numbers were first proposed by Brahmagupta in the 7th century AD although they can be derived by extrapolation from al-Khwarizmi's results
- much later in the 17th century in Europe, negative numbers started to be accepted as solutions of equations
- negative numbers were modelled as directions and debts as a means of justifying the rules for multiplying and dividing negative numbers. However, these rules were not universally accepted because of apparent contradictions.

Preparing for your next review

- You should be prepared to explain how the operations of combining both positive and negative numbers by addition and multiplication can be modelled as movements along the number line.
- You should be aware of the difference between negative and subtracted numbers and be able to discuss why the latter were more acceptable to early users of numbers.
- You should be able to explain how algebraic formulae, involving multiplication of subtracted numbers, may be derived from areas of geometrical figures.
- You should be able to discuss some of the objections to negative numbers that were raised throughout the centuries.
- You should be able to explain some of the limitations in the derivation of rules of multiplication of numbers in the work of, say, Euler and Saunderson.

Searching for the abstract

- Answer the following check questions.

1 What is the difference between a 'negative number' and a 'subtracted number'?

2 List three different ways in which negative numbers may be used in models of physical situations.

3 'If b represents a point on the number line, then b also represents the distance between the point representing 0 and that representing b.'

Is this statement true, or false, or only true under certain circumstances? Give reasons for your answer.

Practice exercises for this chapter are on page 157.

11   Towards a rigorous approach

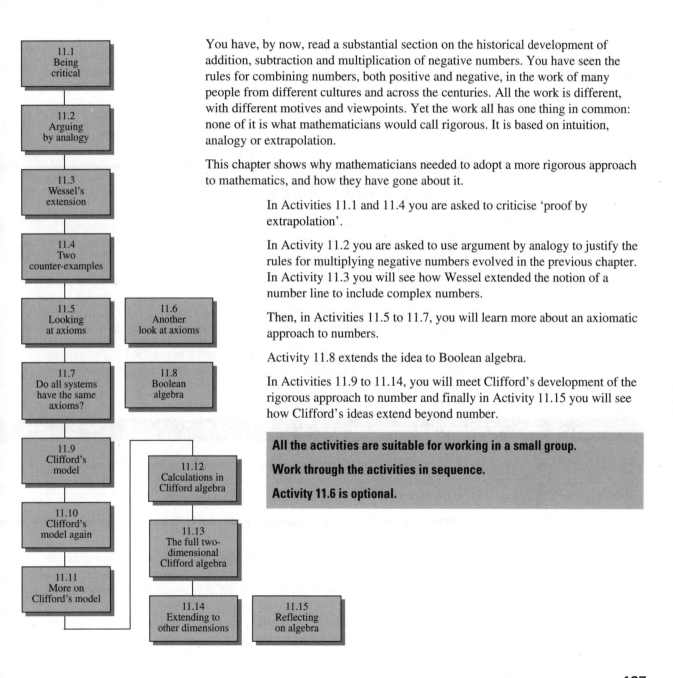

11.1
Being
critical

11.2
Arguing
by analogy

11.3
Wessel's
extension

11.4
Two
counter-examples

11.5
Looking
at axioms

11.6
Another
look at axioms

11.7
Do all systems
have the same
axioms?

11.8
Boolean
algebra

11.9
Clifford's
model

11.12
Calculations in
Clifford algebra

11.10
Clifford's
model again

11.13
The full two-
dimensional
Clifford algebra

11.11
More on
Clifford's model

11.14
Extending to
other dimensions

11.15
Reflecting
on algebra

You have, by now, read a substantial section on the historical development of addition, subtraction and multiplication of negative numbers. You have seen the rules for combining numbers, both positive and negative, in the work of many people from different cultures and across the centuries. All the work is different, with different motives and viewpoints. Yet the work all has one thing in common: none of it is what mathematicians would call rigorous. It is based on intuition, analogy or extrapolation.

This chapter shows why mathematicians needed to adopt a more rigorous approach to mathematics, and how they have gone about it.

In Activities 11.1 and 11.4 you are asked to criticise 'proof by extrapolation'.

In Activity 11.2 you are asked to use argument by analogy to justify the rules for multiplying negative numbers evolved in the previous chapter. In Activity 11.3 you will see how Wessel extended the notion of a number line to include complex numbers.

Then, in Activities 11.5 to 11.7, you will learn more about an axiomatic approach to numbers.

Activity 11.8 extends the idea to Boolean algebra.

In Activities 11.9 to 11.14, you will meet Clifford's development of the rigorous approach to number and finally in Activity 11.15 you will see how Clifford's ideas extend beyond number.

All the activities are suitable for working in a small group.

Work through the activities in sequence.

Activity 11.6 is optional.

Extrapolation

In Activity 10.7, from a rule relating to the product of subtracted numbers you derived the rule for multiplying two negative numbers. This was an example of extrapolation.

Activity 11.1   Being critical

In Chapter 3 of the *Investigating and Proving* unit in Book 4 you were given some advice about how to become a critic of your own and other people's proofs. By adopting this critical role, consider the result you established by extrapolation in Activity 10.7. Can you truly say that you 'proved' your result? Even though each step in your 'proof' may follow logically from the previous one, you may have made assumptions which you did not prove but which are essential to your result.

 1 Examine your result from Activity 10.7 critically, to see if you have proved it. Discuss your thinking with other students, or with your teacher.

Intuition and analogy

An argument based on intuition is one in which you have a reason or reasons to believe that the result is true, but these will not stand up to close scrutiny – they may be based on a hunch rather than on a rigorous proof.

Intuition often depends on being able to visualise the result diagrammatically or pictorially. You can use mathematics to model actual situations and thereby solve practical problems. Equally you can have in mind a practical situation to represent a mathematical system; this can help you to generate results about the system.

For example, if you consider a number line in which positive numbers represent steps to the right and negative ones steps to the left, then putting a minus sign in front of a number is analogous to reversing its direction.

Activity 11.2   Arguing by analogy

1 Use the number line analogy to write an account explaining the following.
a The product of a negative number and a positive number is negative.
b The product of two negative numbers is positive.

It is not necessarily true that intuition and analogy depend on a picture or diagram, or the ability to represent mathematics by an everyday situation. Intuition can be mathematically based. On the basis of your existing mathematical knowledge or experience you may be able to propose another mathematical result in an unrelated, or only semi-related, area. For example, the extract from Euler's text on algebra given in Activity 10.11 displays his mathematical intuition. Intuitively, based on his experience of particular cases, Euler knows that multiplying one number, $-a$, by a second number, $-b$, should not give the same result as multiplying $-a$ by the number b. Euler does not prove that these results are different, but the fact that 2×3 is different from 3×3 makes it seem reasonable.

None of the work to which you have been introduced so far is truly rigorous. If it had been rigorous, then intuitive arguments, analogies and extrapolation would not have been used. All the results would have been firmly established or proved; nothing would have been taken for granted. Mathematics derived from intuition and analogy is often valuable. Much of the mathematics which you use today would not have developed if intuitive arguments had been ruled out. Some of the best mathematics originates as intuitive arguments based on models that help understanding. This is how you, and other mathematicians, are likely to be most creative.

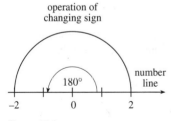

operation of changing sign

number line

180°

−2   0   2

Figure 11.1

However, there comes a time when intuitively based results underpin so much other mathematics that they need to be put on a firmer footing and to be made rigorous. For example, in the mid-18th century new mathematical ideas which relied on the number system were being developed, so it became necessary to establish a rigorous foundation for the number system and of operations based on it.

You can take the study of complex numbers further in the option *Complex numbers and numerical methods.*

You can find a copy of the paper in, for example, *A source book in mathematics,* D E Smith 1959 pp 55–56.

One example of a new mathematical idea based on the number system is the system of complex numbers. A paper written in 1799 by a Norwegian surveyor, Caspar Wessel, shows clearly how the graphical representation of complex numbers is modelled on an extension of the idea of the number line. The new idea is based on the interpretation of a minus sign as producing a reversal of direction along a number line. You can interpret this reversal of direction as a turn through an angle of 180° (Figure 11.1).

Activity 11.3   Wessel's extension

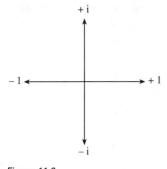

+i

−1 ←——————→ +1

−i

Figure 11.2

It is useful if you already have some knowledge of complex numbers for this activity. However, it is possible for you to try it without this knowledge. Take the imaginary number i to be a line of unit length as shown in Figure 11.2.

Multiplying two negative numbers together involves putting two minus signs in front of the equivalent positive numbers, so involves two turns, each of them through 180°. Adding together the angles of the turning factors involved, namely 180°+180°, gives 360°, or a turn of the positive number through a complete circle. This means that the result is the same as turning through no angle at all, so you remain facing in the positive direction.

You can obtain the same result by thinking of −1 as being a line of length one unit inclined at an angle of 180° to the unit length line +1. Incorporating imaginary numbers into this picture involves extending into a plane. Wessel proposed that the imaginary number i is a line of unit length inclined at 90° to the positive number +1. and that −i is a line of unit length inclined at an angle of 270°.

1 a Multiplying these lines of unit length by adding together the angles that they make with the number +1, show that you can interpret i^2 as a line of unit length inclined to the positive unit at an angle of $90° + 90° = 180°$. How is this consistent with the way the imaginary number i is defined?

Searching for the abstract

b Using Wessel's interpretation of the numbers $+1$, -1, $+i$ and $-i$, work out the results of multiplying all possible pairs of them together. If you already know about complex numbers, then check that your results are as you would expect from your experience of real and imaginary numbers. If you have not met complex numbers before, then explain your results and how you obtained them to your teacher, or to another person who knows about complex numbers, to check your results.

2 If you have not met complex numbers before, or need a reminder about their properties, take a little time to investigate how they behave under multiplication and addition. A complex number has the form $x + iy$, where x and y are real numbers; so examples are $2 + i$, $3.5 - 27i$, $0 + 94i$, and so on.

a Multiplying two complex numbers together is similar to multiplying out two brackets such as $(2 + x)(3 - x)$. What is the outcome of multiplying these two brackets together?

Now try multiplying the brackets $(2 + i)(3 - i)$.

What do you get? Remember that earlier in this activity you showed that $i^2 = -1$.

b Find the outcome of multiplying all pairs of the following complex numbers together: $2 + i$, $3.5 - 27i$, $0 + 94i$.

c Addition of complex numbers is relatively straightforward. The terms involving i are added together separately from the terms without i. For example

$$(2 + i) + (-3 + i) = (2 - 3) + (1 + 1)i$$
$$= -1 + 2i$$

Find the outcome of adding all pairs of the following complex numbers together: $2 + i$, $3.5 - 27i$, $0 + 94i$.

> Wessel's way of multiplying complex numbers together by adding the angles and multiplying the magnitudes is analogous to how you perform multiplication in the modulus–argument representation of complex numbers.

If your answers in Activity 11.3 agree with your experience, then you can say that Wessel's picture provides a faithful representation of the operation of multiplication on complex numbers. Wessel built his interpretation of complex numbers on an extension of a picture for the real numbers; he extended the real number system on the basis of what he assumed was true in the real number system. Such an extension can only be valid if the underlying system is itself on a firm mathematical foundation. Thus the time had come to formalise the rules of the number system to provide a firm mathematical foundation for the extensions that were appearing.

In England, George Peacock (1791–1858) was one of the first to attempt to put algebra, that is, numbers represented by symbols, on a sound logical basis.

At that time, some English mathematicians, including William Frend (1757–1841), still held the view that negative numbers were unacceptable and that trying to prove results relating to them by analogy with debts, was not mathematically valid. Thus Peacock attempted to inject some life into English mathematics by writing two books: *Arithmetical Algebra* in 1842 and *Symbolical Algebra* in 1845. In the first volume he formulated the fundamental laws of arithmetic but for subtracted numbers only. So in that first book $a - b$ has a meaning only if $a > b$. In the second volume he carries all of these rules over to any quantities by the 'principle of the permanence of equivalent forms' which he uses without justification. This is really only a very grand way of describing extrapolation!

1 Find two counter-examples to the 'principle of the permanence of equivalent forms'. That is, find two examples of results which are true for positive numbers but which would not be true if you applied them to negative numbers.

The rules of the number game

The first attempt by Peacock to put negative numbers on a firmer footing did not lead to much success since the same comments could be made of Peacock's work as of Euler's and Saunderson's: he was using extrapolation and analogy. The work of all of them assumed that results held for the negative numbers by analogy with what held for the positives. It was necessary to be even more formal and systematic and to produce rules for combining numbers. The axioms as a whole form a set of 'self-evident' rules from which all the other properties of the numbers can be deduced. The axioms are not proved; rather they are part of the definition of the system which includes ways of combining the numbers. The set of axioms is minimal; this means that properties are not included as axioms if they can be deduced from the other axioms in the set. Basically the axioms are the bare minimum with which you can get by and still be able to deduce all of the properties of the numbers under the usual rules of combination.

The German mathematician Hermann Hankel was the first to produce a logical theory of rational numbers in 1867 but, he had fundamental objections to some aspects of the irrational numbers. The axioms of both the rational and the irrational numbers are attributed to another German, David Hilbert, as late as 1899.

The axioms assume that there are two operations which you can use to combine pairs of numbers. These operations are labelled by + and ×. The effect of each of these operations is to combine two numbers to produce a third number. The third number is not necessarily different from each of the other two. The result of the operation is still a number; it is not some new kind of mathematical object.

1 a The axioms require that there is a particular number which, when combined with any other number using the operation of +, leaves that number unchanged. There is also a number, a different one, which has a similar effect when × is considered instead of +. What are these two numbers?

b If a number is chosen at random, then the axioms require that it is always possible to find another number which, when combined with it using the operation +, gives the number 0. If 2 is the number selected at random, then what is the required number?

c Suppose that you take 2 again and ask for the number which, when combined with 2 using the operation ×, gives 1. What is that number? It you had chosen any other number instead of 2, would you still have been able to satisfy the request? Justify your answer.

d Do you believe that the following statement is true?

'If a number is chosen at random, then the axioms require that it is always possible to find another number which, when combined with the original number using the operation ×, gives 1.'

If you believe the statement is false, then how can you modify it to make it true?

You will not be introduced here to the axioms of the number system in any further detail or be required to deduce from them the familiar and well-used properties of the real numbers. Deductions of this kind can seem tedious since you already use the properties without thinking in your everyday mathematics.

It is also unnecessary, once the axioms have been established and the properties deduced. However it is necessary that someone actually sets up these foundations. Without them, there is a danger that a whole mathematical edifice could crumble due to an overlooked inconsistency.

Axioms are often presented in a very general form. Even though they are describing, for example, the number system with addition and multiplication as the rules of combination, the language may not refer to numbers or to addition and multiplication. It may refer to elements which you might choose to think of as numbers, and the operations for combining the elements might be labelled by using function notation rather than the symbols + and ×.

So the elements may be a, b, ... and an operation to combine them may be labelled by f. In a particular case, you may want to think of f applied to a and b as producing the sum of the numbers a and b, that is, $f(a,b) = a + b$. However this interpretation is not part of the axioms, which are quite abstract.

Activity 11.6   Another look at axioms

This activity is optional.

1 a Write out the requirements of the axioms as given in Activity 11.5 using function notation where $f(a,b) = a + b$ and $g(a,b) = a \times b$.

b Is it true that $f(a,b) = f(b,a)$ and $g(a,b) = g(b,a)$? Justify your answers.

Other developments in the 19th century

Cayley was one of the first people to study matrices – see the *Modelling with matrices* unit in Book 4.

The 19th century was an active period for European mathematics, in England and Ireland in particular. You have already read about Peacock; other active mathematicians in England at the time were Augustus De Morgan (1806–1871), Arthur Cayley (1821–1895); William Rowan Hamilton (1805–1865) was from Trinity College in Dublin. You will be hearing about some of their attempts to extend the number system.

De Morgan was very keen to describe the number system in an abstract way, and to keep numbers themselves out of an axiomatic description of the number system.

There was a major implication of this approach. All systems which obey the axioms also obey any rules derived from the axioms. This is true of the real and complex numbers, both of which satisfy the axioms. However, Hamilton was inventing an extension (which he called quaternions) of complex numbers. You will see later that quaternions do not satisfy the same rules as numbers.

Activity 11.7   Do all systems have the same axioms?

1   Check that De Morgan's view is correct at least as far as the real and complex numbers are concerned. That is, check that the axioms as specified in Activity 11.5, question **1a,** for combining elements under + and × are the same for both the real and the complex number systems.

2   Also, given any real number a, the axioms require you to be able to specify another number b such that $a + b = 0$. If $a \neq 0$, the axioms also require you to be able to specify another number c such that $a \times c = 1$. If a is the complex number $2 + i$, find b and c.

George Boole (1815–1864) had no formal mathematical education. He became a primary school teacher and during that time taught himself more mathematics and became interested in a debate over logic which was taking place. As a result, in 1847 he published a short work on logic which was recognised by De Morgan as being revolutionary. Nevertheless it did not receive much recognition at the time although it probably enabled Boole to get a position teaching at Queen's College in Cork. Seven years later he produced the classic *Investigation into the laws of thought*, which established the algebra of logic, or the algebra of sets, now known as Boolean algebra. In other words, Boole set up the rules by which the elements in his sets, or in his logic, might be combined.

Activity 11.8   Boolean algebra

1   a   Find out as much as you can about Boole and Boolean algebra and make notes as you do your research.

b   As you do your research, decide for yourself whether the elements of Boole's algebra satisfy the same rules of combination as the real numbers.

c   Use your notes to prepare and make a presentation to your fellow students on Boole and Boolean algebra. As part of your presentation, include your answer to part **b**. Remember that your fellow students may not know anything about axioms, so make your presentation in a way which is easy for them to understand.

William Kingdon Clifford (1845–1879) was a graduate of Trinity College, Cambridge. He was renowned for his teaching, and for his ability to visualise mathematical ideas and bring them to life. Clifford almost always uses models in which algebraic quantities can be visualised geometrically or diagrammatically.

For the system of real numbers, Clifford takes a step to the right of length 1 to be the basic element from which everything else is built. Multiplying this basic step by a positive number, another step, gives a longer step. Other operations on this step are those labelled by + and –, which represent the operations of 'retaining' or 'reversing' the direction of the step.

Here is Clifford's own explanation.

> In the equation
> $$2 \times (+3) = +6$$
> the last term on each side is a step, the first is an operator and the equation means by doubling I can turn a step 3 to the right into a step 6 to the right.
>
> Now what operator is required to turn the step -3 to the left into the step $+6$ to the right?
>
> First we reverse the step by an operator which we will call r, $\{r(-3)\} = +3$: thus it becomes +3. Now double it, and the whole operation is written,
> $$2r(-3) = +6,$$
> so the required operator is $2r$, which means reverse and then multiply by 2.
>
> But we may change the order of the process, viz. double and then reverse and we get the same result
> $$r2(-3) = +6.$$
> … Let $k3$ mean triple without reversing. And let us suppose any step taken, tripled, reversed, doubled and reversed again. The two reversals will clearly destroy each other and give 'no reversal' or k, and we shall have our step sextupled without reversal. This may be written in an equation like the last. And in the same way we have two others in which the direction is reversed:
>
> | I | $r2(-3) = +6$ | $r2(+3) = -6$ |
> | II | $r2(r3) = k6$ | $r2(k3) = r6$: |
>
> and we are led to assign a new meaning to the symbols + and –; we may use them instead of k and r respectively.

Activity 11.9 Clifford's model

1 Clifford interprets all real numbers and the operations on them by taking as the basic building blocks a step of length 1 to the right and the operations of +, – and $\times$.
a Describe how you would obtain from basic building blocks a step of length 12 to the left.
b Describe the process of taking a step of 2 to the left, tripling it, reversing its direction and then doubling it. Write the process as an algebraic expression.

Describe other ways in which you can start with the same step and achieve the same outcome. By analogy, what can you deduce about the rules for combining numbers using the operations of × and +?

2 You can think of the basic building block of a step of length 1 as being a step corresponding to some standard size, like a metre length. You can make a smaller step than the basic one by multiplying it by a positive number a, where $a < 1$.

a Describe how you would obtain a step to the left which has a length of one-half of the standard step.

b If you performed this step three successive times and then reversed its direction, what would be the outcome? How might you express this process in symbols?

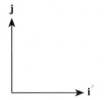

Figure 11.3

You can extend Clifford's picture to two dimensions. The basic element is still a step but it can come in different varieties because there are many more directions in which the step can be taken in the plane compared with the line. However, all steps in the plane can be built up from two basic vectors together with the operations of + and – as described above. The two basic vectors **i** and **j** are those with which you are already familiar. Numbers, called scalars, are the size of the steps. Then, for example, you can interpret multiplying a vector by a positive number a as changing the size of the step while leaving its direction unchanged. If $a > 1$ then the size increases, whereas if $a < 1$ then the size decreases.

Activity 11.10   Clifford's model again

1 Referring to Figure 11.3, draw diagrams to represent in Clifford's model the outcome of the following combinations of steps with other steps and with scalars.

a $-2 \times \mathbf{i}$ **b** $\mathbf{i} + \mathbf{j}$ ◈ **c** $-\left(\frac{1}{3} \times \mathbf{i}\right) - \left(\frac{3}{2} \times \mathbf{j}\right)$

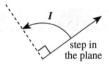

Figure 11.4

By multiplying **i** by the scalar −1 you can turn it into −**i**; similarly you can turn **j** into −**j**. Thus, multiplication of a step by a negative number allows you to move up and down a line in the plane, but it does not allow you to convert one line into another. For example, to turn **i** into **j** you need a turning operator; call this turning operator I. Figure 11.4 shows the way I operates.

When you apply the operator I to a step, it turns through 90° in an anti-clockwise direction. In particular, it will turn **i** into **j** and **j** into −**i**. Using symbols you can write this as

$$I \times \mathbf{i} = \mathbf{j} \text{ and } I \times \mathbf{j} = -\mathbf{i}$$

Here the multiplication symbol × is being used to represent the application of the turning operator to the step.

Activity 11.11   More on Clifford's model

1 Use Clifford's interpretation of I, **i** and **j** to show that the following statements are true.

a $I \times I \times \mathbf{i} = -\mathbf{i}$ **b** $I \times I \times \mathbf{j} = -\mathbf{j}$ **c** $I^2 = -1$

As a result of the property in part **c** you can see that I behaves like the imaginary number i. It is then natural to form composite numbers like $1+I$, $2-I$, and so on.

2 What is the outcome when these composite numbers are allowed to act on **i** and **j**? That is, what is the result of the following algebraic operations?

$$(1+I)\times \mathbf{i} \qquad (1+I)\times \mathbf{j} \qquad (2-I)\times \mathbf{i} \qquad (2-I)\times \mathbf{j}$$

Suggest a geometrical interpretation of the composite numbers $1+I$ and $2-I$.

3 What is the outcome of multiplying the turning operator by a number? Make a suggestion by considering $2\times I$ applied to the step **i**. Also consider the application of $1+\left(\frac{2}{3}\times I\right)$ to **i**. Investigate other possibilities. What are your observations?

Thus Clifford's interpretation of the real numbers in two dimensions produces complex numbers. It can also produce something quite new. Suppose that you extend the definition of $\times$ to multiply vectors. Assume $\times$ satisfies the properties:

1 $\mathbf{i}\times\mathbf{i}=1$

2 $\mathbf{j}\times\mathbf{j}=1$

3 $\mathbf{i}\times\mathbf{j}\neq 0$ but instead defines a new object which is neither a scalar nor a vector.

You can think of properties 1 to 3 as axioms for multiplying vectors.

Apart from property 3, Clifford's method of multiplying vectors is similar to the scalar product. The operation $\times$ is the same as the scalar product when you multiply a vector by itself. Otherwise $\times$ is quite different; when you multiply two different vectors **a** and **b**, you obtain a new quantity, labelled for now by $\mathbf{a}\times\mathbf{b}$.

You will investigate the implications of property 3 after doing some algebra in Activity 11.12.

Activity 11.12   Calculations in Clifford algebra

1 If $\mathbf{a}=2\mathbf{i}-3\mathbf{j}$ and $\mathbf{b}=-\mathbf{i}+5\mathbf{j}$, show that

$$\mathbf{a}\times\mathbf{b}=-17+10(\mathbf{i}\times\mathbf{j})+3(\mathbf{j}\times\mathbf{i})$$

2 a By using $I\times\mathbf{i}=\mathbf{j}$ and $I\times\mathbf{j}=-\mathbf{i}$ and properties 1 and 2, show that $I=-\mathbf{i}\times\mathbf{j}$ and $I=\mathbf{j}\times\mathbf{i}$.

b Have you made any other assumptions in deducing these results? If so, what?

c Deduce that $\mathbf{j}\times\mathbf{i}=-\mathbf{i}\times\mathbf{j}$.

3 If $\mathbf{a}=-6\mathbf{i}-\mathbf{j}$ and $\mathbf{b}=-\mathbf{i}-\mathbf{j}$, show that $\mathbf{a}\times\mathbf{b}=7+(5\times\mathbf{i}\times\mathbf{j})$ and that $\mathbf{b}\times\mathbf{a}=7-(5\times\mathbf{i}\times\mathbf{j})$.

As a result of your work on Activity 11.12, you can see that you have found something completely new. When you combine together real numbers by + or by $\times$, the order in which you combine them makes no difference to the result. Thus $-2\times6=6\times-2$ and $6+3=3+6$.

The operations of + and × on the real numbers are said to be **commutative**. With the operation × on vectors, you have found an operation which is not commutative; that is, the order in which you combine the vectors does affect the result.

Consequently, the rules which govern the combination of objects by × would not look the same as the rules governing the combination of real numbers under either + or ×. So De Morgan was wrong; not all systems need necessarily obey the same rules. You can invent a new system, an extension of the number system, with different rules from the number system.

Activity 11.13   The full two-dimensional Clifford algebra

Activity 11.12, question **2**, shows that the turning operator *I* which turns one step into another is a product of the two basic vectors **i** and **j** in Clifford's model of two dimensions. It is called the basic bivector in the model.

1 **a** Investigate combinations of any number of vectors in two dimensions using Clifford's operation ×.
b Also investigate combinations of vectors and bivectors, bivectors and bivectors, numbers and bivectors. What types of objects are you able to produce? Can you come up with any rules to indicate the type of object that you will obtain from a product of particular objects?
c What are the basic building blocks in Clifford's two-dimensional model?

Clifford's three kinds of elements in two dimensions,
- scalars (that is, numbers or the size and direction of steps)
- vectors **i** and **j** (that is, steps)
- the bivector **j** × **i**

form the basic building blocks of the most general algebraic system in two dimensions, which is called a Clifford algebra. In a Clifford algebra you are allowed to combine together any of these elements.

You have met the idea of extension on several occasions now. Wessel extended the real number system to two dimensions in order to define the system of complex numbers with similar rules of addition and multiplication to those of the real numbers. His extension was based on a particular model of the number line and operations on it. Clifford had a different model of the real numbers which, when extended, produced his algebra of geometrical objects in two dimensions which have different rules for addition and multiplication from the real numbers.

Activity 11.14   Extending to other dimensions

For further details see Chapter 18 of the *Space and vectors* unit of Book 3.

Clifford's approach can be further extended to three dimensions. Steps are built up from three basic vectors **i**, **j** and **k**.

In three dimensions you can define three turning operators *I*, *J* and *K* as shown in Figure 11.5.

Searching for the abstract

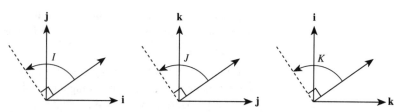

Figure 11.5

1 By analogy with the equations considered in two dimensions, find all the equations satisfied by the vectors and the turning operations in three dimensions. Notice that the result of $I \times \mathbf{k}$ cannot be defined as a vector, since the operator I can only act on vectors in the plane containing $\mathbf{i}$ and $\mathbf{j}$ to produce other vectors in the same plane. In general $I \times \mathbf{k}$ will produce a new object. Similar comments apply to $J \times \mathbf{i}$ and $K \times \mathbf{j}$.

2 Deduce the following.

a $I^2 = J^2 = K^2 = -1$

b $I \times J \times K = -1$

c $I \times J = -J \times I = K$

These three results are the defining relations of quaternions, the name given to the three turning operators I, J and K, which were invented by Hamilton.

Clifford's algebra in three dimensions is formed from the basic building blocks: the scalars; vectors; bivectors or the turning operators $\mathbf{i} \times \mathbf{j}$, $\mathbf{i} \times \mathbf{k}$, $\mathbf{j} \times \mathbf{k}$; and the trivector $\mathbf{i} \times \mathbf{j} \times \mathbf{k}$. All other distances, steps and turns can be produced from these basic building blocks.

So far, all your extensions of Clifford's original approach have been to spaces which you can visualise. In two and three dimensions, you can draw diagrams to represent the building blocks. Just because you cannot visualise, or draw pictures in, dimensions any higher than three does not mean that there are no more Clifford algebras to discover.

You can produce a four-dimensional algebra by generalising the three-dimensional Clifford algebra. Instead of using three basic vectors, you can use four. There will then be six turning operators, and so on.

> How does your knowledge of permutations help you to deduce that there must be six turning operators when there are four vectors?

A complete generalisation of the Clifford algebra is obtained when you do not specify a particular number of dimensions – and hence the number of basic steps – instead you specify how to construct the algebra for n basic vectors. The method of construction would work satisfactorily for any value of n, 1, 2, 3, 27, 102, or whatever you choose.

Activity 11.15 Reflecting on algebra

1 Find an example from Books 1 to 5, or from your wider experience, where the elements of a collection of objects other than numbers are being combined by two different operations, rather like addition and multiplication of real numbers, although these operations may not be labelled by + and ×. Write down the rules for combining these objects. Translate these rules into a new language by consistently replacing the symbols used for your two operations by the symbols + and ×. If your objects were represented by numbers, would these translated rules look the same as the rules for combining real numbers by addition and multiplication? If so, then your collection of objects satisfies the same axioms as the real numbers. If not, then the axioms for the two systems are different; in which case, how do they differ?

Charting the progress and status of negative numbers through history highlights some interesting points concerning the development of any piece of mathematics. It has shown how many results are first generated by analogy, intuition and extrapolation, and do not necessarily develop a rigorous foundation until later.

However, once the rigour has been established, extensions and generalisations can be produced in which the rules governing the combination of numbers by + and × are used to represent the combination of new objects. These rules then acquire the status of the rules governing an algebraic system, since they no longer just apply to the original numbers. And, potentially, this algebraic system allows the possibility of generalisation to an abstract structure, which exists in spaces which you cannot necessarily visualise.

Reflecting on Chapter 11

What you should know
- that mathematical results are conjectured using various methods: extrapolation, analogy, intuition; and that none of these methods is necessarily rigorous
- that complex numbers can be described as an extension of the real number system, and that they may be modelled in the plane
- that the properties of negative numbers do not follow the 'principle of permanence of equivalent forms'
- that it is possible to model a positive number as a step along the number line and that multiplication of the number by −1 is analogous to reversing its direction
- that it is possible to extend this model of the real numbers into the plane to produce a different algebraic system from the complex numbers (this is Clifford's model)
- that the unit vectors can be modelled as basic steps at right angles to one another in both two and three dimensions, and that it is possible to define an operator which turns one basic step into the other
- that it is possible to combine numbers, steps and turning operators, using an operation analogous to multiplication, but which has different properties from multiplication of numbers
- that systems, which can be visualised and represented by geometrical objects, can be extended and generalised to more abstract situations, for example, to higher dimensions, which can no longer be represented pictorially but which are equally valid mathematically.

Preparing for your next review

● You should be able to describe examples of results that have been conjectured through each of the methods of extrapolation, analogy and intuition.

● You should be able to explain the properties of the imaginary number i by modelling it as a line of unit length inclined at 90° to the unit length line +1.

● You should be able to describe and to justify examples of systems which have the same properties (axioms) as real numbers and those with different properties.

● You should be ready to explain how the system of real numbers can be extended in several ways to describe new mathematical systems.

● You should be able to explain, using diagrams, the basic elements of Clifford's model in two and three dimensions and to justify their properties.

● You should be able to calculate the results of combining together various elements in Clifford's model using addition and multiplication.

● You should be aware of the potential to generalise systems such as the real numbers or complex numbers.

● Answer the following check questions.

1 By referring to the mathematics in Books 1–5, list some mathematical ideas and systems which you now see as extensions and/or generalisations of the real numbers. Give reasons for your answers by referring to the rules for combining the elements in your various examples.

2 What is the 'principle of permanence of equivalent forms' as proposed by George Peacock? Give an example which supports the principle when the positive numbers are extended to include the negative ones.

3 Give an example of a result in number systems which was originally proposed by extrapolation from a previously known result.

4 Write down two different geometrical interpretations of the imaginary number i and briefly say why each of them is consistent with $i^2 = -1$.

5 How do quaternions differ from rational numbers in their rule of multiplication? Write down the rules for multiplying the quaternion units I, J and K.

6 By drawing diagrams, explain why you need three different kinds of basic turning operators in three dimensions. How many would you need in five dimensions?

Practice exercises for this chapter are on page 157.

12 Summaries and exercises

1 Introduction to the Babylonians

Chapter summary

- general background (whole chapter).

Practice exercises

There are no practice exercises for this chapter.

2 Babylonian mathematics

Chapter summary

- an introduction to the Babylonian number system (Activities 2.1 to 2.5)
- Babylonian arithmetic (Activity 2.6)
- the Babylonian method for finding square roots (Activities 2.7 to 2.9)
- solution of quadratic equations (Activities 2.10 and 2.11)
- Babylonian geometry (Activities 2.12 and 2.13).

Practice exercises

1 Look at Figure 12.1 and answer the following questions.

a Using your calculator or otherwise, convert 42; 25,35 and 1; 24,51,10 into decimal numbers. Give your answers to at least seven significant figures accuracy.
b To which number is 1; 24,51,10 an approximation?
c Describe a geometric algorithm to estimate square roots which the Babylonians may have used to calculate this approximation.
d Write the algorithm to calculate $\sqrt{N}$ as you would implement it on a calculator.
e Implement your algorithm on your calculator, starting with 2, to find an approximation to $\sqrt{7}$. Write down the first five iterations.

2 Comment on the claim that the Babylonians knew about Pythagoras's theorem 1200 years before Pythagoras lived.

3 Find a number which when added to its reciprocal becomes 2; 0,0,33,20.

4 The following quotation is adapted from *The history of mathematics: a reader*, edited by J Fauvel and J Gray, pages 42–43.

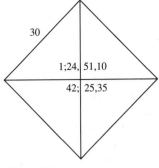

30

1;24, 51,10

42; 25,35

Figure 12.1
Transcription of Figure 2.8b

> Two texts which call for comment here represent versions of the same exercise, and they are significant, not only because of the abstract interest in numerical relations indicated by the enormous numbers involved, but also because they concern the problem of irregular numbers.
>
> Number 671 is written by a bungler who succeeded in making half a dozen writing errors in as many lines ...

Jestin Number 50	Jestin Number 671	Translation of Number 50
še guru$_7$:1	(rev) še sila$_3$ 7$^!$ guru$_7$	The grain (is) 1 silo.
sila$_3$ 7		7 sila (7 liters)
lú: 1 šu ba-ti	lú: 1 šu ba-ti	each man received.
lú-bi	(obv) guruš	Its men:
45,42,51	45,36,0 (written	45,42,51.
	on three lines)	
še sila$_3$:3 šu $^?$-tag$_4$$^?$		3 sila of grain (remaining)

> A silo (guru) in this period contained 40,0 gur, each of which contained 8,0 sila. The correct answer seems to have been obtained by the following process:
>
> (1) 5,20,0,0 times 0; 8,34,17,8 = 45,42,51; 22,40
>
> (2) 45,42,51 times 7 = 5,19,59,57
>
> (3) 5,20,0,0 − 5,19,59,57 = 3

a   Explain in your own words what problem the students have been asked to solve.
b   Explain where the number 5,20,0,0 comes from.
c   Explain step (1) in your own words.
d   Verify that 45,42,51 times 7 equals 5,19,59,57.
e   What mathematical error has the pupil apparently made in Jestin 671?

5   Give a geometric and an algebraic description of the Babylonian method of solving an equation of the form $x^2 + ax = b$.

3   An introduction to Euclid

Chapter summary

- how to use the Greek rules to carry out constructions for drawing a perpendicular, bisecting an angle, bisecting a line segment, drawing a perpendicular to a line segment (Activities 3.1 and 3.2)
- how to construct a square equal in area to any given polygon (Activity 3.3)
- the significance of the three classical problems of ancient Greece (Activity 3.4)
- some Euclidean proofs to work through (Activities 3.5 and 3.6)
- how Euclid's approach established an axiomatic basis for geometry (Activities 3.7 and 3.8).

Practice exercises

1   Given a line segment AB and a straight line l, use Euclid's rules to construct points C and D on l such that CD = AB.

Consider the following different cases.

a   *l* intersects AB.

b   *l* intersects an extension of AB.

c   [Harder] *l* is parallel to AB.

2   Given a line segment AB and a point P on AB, use Euclid's rules to construct on AB a point Q such that $AQ = BP$.

3   Explain how to construct a square equal in area to a given triangle.

4   More Greek mathematics

Chapter summary

- the major contributions and the background to the lives of
  - Euclid (Chapter 3 and Activities 4.1 to 4.3)
  - Archimedes (Activities 4.6 and 4.7)
  - Apollonius (Activities 4.8 and 4.9)
  - Hypatia (text at the end of Chapter 4)
- how to interpret Euclidean theorems of geometric algebra in modern algebraic notation (Activity 4.1)
- how to use the Euclidean algorithm to find the highest common factor (Activity 4.2)
- how the ancient Greeks proved the infinity of primes (Activity 4.3)
- the meaning of 'incommensurability' and the significance of its discovery to the ancient Greeks (Activity 4.4)
- an example of how the Greeks continued to take an interest in numerical methods: Heron's algorithm for finding square roots (Activity 4.5)
- how to use and interpret Archimedes's method for estimating π (Activity 4.6)
- how Archimedes found the area under a parabolic arc (Activity 4.7)
- how Apollonius defined and named the conic sections (Activities 4.8 and 4.9).

Practice exercises

1   Here is a proposition from Euclid.

> If there are two straight lines, and one of them can be cut into any number of segments, the rectangle contained by the two straight lines is equal to the rectangles contained by the uncut straight line and each of the segments.

a   Construct a diagram to represent the proposition.

b   Make an algebraic statement equivalent to the proposition.

2   Use the Euclidean algorithm to find the greatest common divisor of 2689 and 4001.

3   Prove proposition 14 from *Book VIII* of Euclid's *Elements* that, if a^2 measures b^2, then a measures b, and, conversely, that if a measures b, then a^2 measures b^2.

4   Write brief notes on Archimedes's contribution to mathematics.

5 Arab mathematics

Chapter summary

- some inheritance problems from Islamic law (Activity 5.1).
- how Arabic work on trigonometry built on the work of Ptolemy to compile trigonometric tables (Activities 5.2 and 5.3)
- examples of problems from the East worked on by Arab mathematicians (Activity 5.4)
- how Thabit ibn Qurra investigated amicable numbers (Activity 5.5)
- some Chinese methods for solving equations which were available to the Arabs (Activities 5.6 and 5.7)
- the origins of the word 'algebra' (text before Activity 5.8)
- the six types of quadratic equation solved algorithmically, with geometric justification, by al-Khwarizmi (text and Activity 5.8)
- some of the 19 types of cubic solved by Omar Khayyam (text and Activity 5.9)
- methods for finding local maxima and minima to help find the solutions of a cubic (Activity 5.10).

Practice exercises

1 A particular equation which was considered by al-Khwarizmi can be translated as follows.

> one square and ten roots of the same equal thirty-nine.

a Express this equation in a modern algebraic form.
b Draw a square ABCD with side AB representing the 'root' of the equation. Extend AB to E and AD to F where $BE = DF = 5$, and complete the larger square on side AE. Extend the lines BC and DC to meet the larger square. Shade the areas representing the 'one square and ten roots of the same' in al-Khwarizmi's problem.
c Hence find the positive solution to the equation.

2 Explain why al-Khwarizmi's classification of quadratic equations does not include the case or cases in which squares and roots and numbers equal zero.

3 Which of the Arabic numerals look most like those used in English texts today? What advantages and disadvantages can you see in the Arabic versions?

4 Use al-Khwarizmi's method to solve $x^2 + 12x = 64$.

5 Use Horner's method to solve $x^2 + 7x = 60750$.

6 Write brief notes on Omar Khayyam's contribution to mathematics.

7 Draw on your studies of Babylonian and Arab mathematics to write an account of two different approaches to problems which lead to quadratic equations.

6 The approach of Descartes

Chapter summary

- a review of the legacy of Greek geometrical methods (Activities 6.2 and 6.3)

- how Viète's law of homogeneity restricted the development of algebra (Activities 6.4 and 6.5)
- how Descartes's method represented a new approach to the use of algebraic notation (Activities 6.6 to 6.11)
- how to apply Descartes's method to van Schooten's problem (Activity 6.12)
- an algebraic analysis of the problem of trisecting an angle (Activity 6.13).

Practice exercises

The practice exercises for this chapter are combined with those in Chapter 8.

7   Constructing algebraic solutions

Chapter summary

- how to translate an algebraic analysis into a geometric construction (Activity 7.1)
- how Descartes developed constructions requiring circles (Activities 7.2 to 7.6)
- how Descartes's method can lead to solutions which are curves (Activities 7.7 to 7.9)
- Descartes's discussion of the nature of curves (Activity 7.10 and preceding text)
- analysing curves produced using construction instruments (Activities 7.11 to 7.13)
- Descartes's discussion of acceptable curves (Activity 7.14)
- Descartes's geometrical construction of the trisection of an angle (Activity 7.15).

Practice exercises

The practice exercises for this chapter are combined with those in Chapter 8.

8   An overview of La Géométrie

Chapter summary

- the importance of Descartes's work seen so far (Activity 8.1)
- how Descartes classified curves (Activities 8.2 to 8.4)
- how Descartes attempted to standardise equations (Activity 8.5)
- how Descartes treated negative solutions (Activity 8.6)
- how to use Descartes's rule of signs (Activities 8.7 and 8.8)
- how Descartes constructed a normal to a curve (Activity 8.9)
- the contribution of Descartes to analytical geometry (Activity 8.10, and surrounding text)
- an example of the work of Fermat in analytical geometry (Activity 8.11).

Practice exercises

1   Write brief notes on Descartes's contribution to mathematics.

2   Construct Descartes's graphical solution to the cubic equation $x^3 = -4x + 16$,

using the circle $(x-8)^2 + \left(y + \frac{3}{2}\right)^2 = 8^2 + \left(\frac{3}{2}\right)^2$ and the parabola $x^2 = y$. Explain why the intersection of this circle and the parabola provides the solution.

3   Use the method of Descartes to solve the equation $y^2 = -ay + b^2$.

9   The beginnings of calculus

Chapter summary

- how Euclid defined a tangent to a circle (Activity 9.1)
- how Archimedes constructed a tangent to a spiral (Activity 9.2)
- how Fermat found maxima and minima, and how he used this in his general method to construct tangents to a curve (Activities 9.3 and 9.4)
- how Descartes's definition of a tangent, and how his method for constructing tangents differed from Fermat's definition and method (Activity 9.5)
- the development of integral notation (Activities 9.6 and 9.7)
- Leibniz: gradient and area (Activity 9.8)
- Newton and differentiation (Activity 9.9).

Practice exercises

1   Write brief notes on Leibniz's contribution to mathematics.

2   What three long-standing classes of problems were solved by the introduction of the calculus? Explain why calculus was important in the solution of these problems.

10   'Two minuses make a plus'

Chapter summary

- how to visualise addition and multiplication on a number line (Activity 10.1)
- subtracted numbers (Activity 10.2)
- an example of subtracted numbers arising in ancient China (Activity 10.3)
- Chinese rules for combining negative numbers (Activity 10.4)
- a geometric approach to multiplying negative numbers using Indian, Arab and Greek sources (Activities 10.5 to 10.7)
- thinking about why there is a gap in development (Activity 10.8)
- some particular difficulties with negative numbers (Activities 10.9 and 10.10)
- a debate based on source material (Activity 10.11).

Practice exercises

The practice exercises for this chapter are combined with those in Chapter 11.

11   Towards a rigorous approach

Chapter summary

- thinking about the validity of proof by extrapolation (Activity 11.1 and 11.4)
- using analogy to justify the rules for combining negative numbers (Activity 11.2)

- extending the number line to include complex numbers (Activity 11.3)
- looking at an axiomatic approach (Activities 11.5 to 11.7)
- Boolean algebra as an example of an axiomatic system (Activity 11.8)
- exploring Clifford algebra (Activities 11.9 to 11.14)
- generalising beyond numbers (Activity 11.15).

Practice exercises

1 In his *Arithmetica infinitorum*, John Wallis gives his view that a number larger than infinity can be found.

> a ratio greater than infinity such as a positive number may be supposed to have to a negative number

a Investigate the sequence of positive ratios $\frac{1}{2} \ \frac{1}{3} \ \frac{1}{4} \ \frac{1}{5}$. This is a sequence of the form $\frac{1}{n}$ with n increasing. In numerical terms is the sequence descending or ascending?

b Now investigate the sequence of ratios $\frac{1}{n}$ where n is any positive number. What can you say about the sequence as n tends towards zero? Draw a graph using n as the horizontal axis.

c By extending the sequence to negative values of n, again with n decreasing, can you explain why John Wallis may have put forward the view expressed?

d What is wrong with Wallis's argument?

2 In the book *Mathematical thought from ancient to modern times*, Kline assigns to Augustus de Morgan the following:

> The imaginary expression $\sqrt{(-a)}$ and the negative expression $-b$ have this resemblance, that either of them occurring as the solution of a problem indicates some inconsistency or absurdity. De Morgan illustrated this by means of a problem. A father is 56; his son is 29. When will the father be twice as old as the son?

a Solve de Morgan's problem by setting up an equation to find out how many years from now the father will be twice as old as the son.

b In what way does your solution support de Morgan's claim?

c How do you interpret your solution and what assumptions are you making about numbers?

d How can de Morgan's original question be phrased so that the solution no longer appears absurd?

3 Discuss critically various contradictory beliefs which have been held about negative numbers throughout the centuries and throughout the world.

13 Hints

Activity 2.3, page 15

2 First exclude the two 'exceptional' numbers in the right-hand column.

3 What is a half in decimal notation, and in Babylonian?

Activity 2.5, page 17

3 Writing a fraction in sexagesimal notation is comparable with the following decimal conversion.

$$\frac{1}{4} = \frac{10 \times \frac{1}{4}}{10} = \frac{2\frac{1}{2}}{10} = \frac{2}{10} + \frac{\frac{1}{2}}{10} = \frac{2}{10} + \frac{10 \times \frac{1}{2}}{10^2} = \frac{2}{10} + \frac{5}{10^2} = 0.25$$

Activity 2.11, page 24

2 $x - y = 7$ from line 1; $\frac{1}{2}(x - y) = \frac{1}{2} \times 7$ from lines 3 to 5.

Activity 3.2, page 32

4 b A median of a triangle is a straight line from one vertex of a triangle to the mid-point of the opposite side.

Activity 3.3, page 34

1 g Draw an arc of a circle, with centre H and radius HI. Extend GH to intersect this circle outside GH at N. Then GN is a line segment divided at H into lengths a and b. Now you need to construct a right-angled triangle with hypotenuse GN. Use the fact that all vertices C of the right-angled triangles ABC, of which AB is the hypotenuse, form a semi-circle with diameter AB.

Activity 4.7, page 49

1 b Think about the coefficient of x^2 in the parabola.

c Think about the difference between calculating QM given the x-coordinates of A and B, and finding the lengths UV and HK given the x-coordinates of P, M, R.

Activity 4.9, page 50

1 b Notice that the section MPN is parallel to the base BCE of the cone, and so is a circle. What does this tell you about the triangle MPN? How does this fact help you to use the idea of a mean proportional?

c Explain why triangles AHF and FLM are similar. How does this lead to the result $LM = FL \times \dfrac{HF}{AH}$?

2 a As MN moves the triangles FLM are always similar to each other, and the triangles JLN are always similar to each other.

Activity 5.4, page 60

3 You need to take some care to search effectively for solutions of the two equations $A + B + C + D = 100$ and $189A + 225B + 245C + 945D = 31500$.
- Notice that A must be a multiple of 5, B must be a multiple of 7, and then that C must be a multiple of 9.
- Notice also that the largest possible value of D is 33.
- Make sure also that when you search you reject any cases when $A + B + C + D > 100$ or $189A + 225B + 245C + 945D > 31500$.

Even so, the Fowls algorithm takes 102 minutes to run on one type of calculator.

Fowls

Input None.

```
0 → N
0 → A
0 → B
0 → C
0 → D
while A + B + C + D ≤ 100 and 189A + 225B + 245C + 945D ≤ 31500 do
    for A = 0 to 100, step 5
        for B = 0 to 100, step 7
            for C = 0 to 100, step 9
                for D = 0 to 33, step 1
                    if A + B + C + D = 100 and 189A + 225B + 245C + 945D = 31500
                    then
                            display A
                            display B
                            display C
                            display D
                            N + 1 → N
                            display N
                    endif
                next
            next
        next
    next
endwhile
```

Output A list of solutions, and the number of solutions.

Activity 6.13, page 85

4 First use triangles ORM and OQT to prove that $QR = \dfrac{3z - q}{z}$. Then find another expression for QR in terms of q and z. Finally equate the two expressions for QR.

13 Hints

Activity 7.3, page 89

1 Write MP and MO in the forms $MP = MN - NL$ and $MO = MN + NL$.

Activity 9.8, page 120

3 What is the size of the difference between the areas?

Activity 10.1, page 125

2 b Remember that 2×3 is the same as $0 + 3 + 3$. Think about $-3 + (-3)$. How can you can write this in multiplication notation? Does this help you extend further your number line model? You might have difficulty extending this model to include examples such as -4×2 and -4×-2. If your model can interpret these types of examples, discuss it with a friend or your teacher.

Activity 10.2, page 125

1 You are much more likely to come across subtracted numbers, as opposed to free standing negative numbers, in the modelling of practical problems. Can you see how these might have made them more acceptable? Can you think of some examples where subtracted numbers arise in problems relating to everyday situations? Would free standing negative numbers arise in the same situations in a meaningful way?

Activity 10.3, page 126

1 Set out the problem as

$$-5x + 6y + 8z = 290$$
$$3x - 9y + 3z = 0$$
$$-13x + 2y + 5z = -580$$

where x, y and z are the prices of cows, sheep and pigs respectively. Solve these equations for x, y and z.

Activity 10.5, page 127

1 The area of the rectangle with sides $(a - b)$ and $(c - d)$ is the same as the area of the large rectangle with sides a and c, provided that you subtract off the area of the L-shaped border. Write this statement as an equation.

Activity 10.6, page 128

1 a The first part of the extract explains about the construction of the geometrical figure shown in the diagram. The result of the proposition is actually described in the section of the extract entitled *Demonstration*. You should concentrate on this part. Try to identify the squares and rectangles referred to in this section and then describe in your own words the relationship between them. Note that the extract refers to squares and rectangles by just two points, those at opposite ends of one of the diagonals of the square or rectangle.

Introduce letters to represent the lengths of the sides of the squares, then you should be able to write down an algebraic formula which is equivalent to the result.

Activity 10.7, page 129

1 Make a special choice of the numbers *a* and *c* which reduces the equation to one which just has two negative numbers multiplied together on the left-hand side. Having made this choice, substitute it in both sides of al-Khwarizmi's equation to obtain the rule for multiplying negative numbers together.

2 Follow the same sort of argument here as applied in question **1** except that you are not necessarily restricting yourself to choices of *a* and *c*. Remember that division by a number *e* is the same as the operation of multiplication with $\frac{1}{e}$.

Activity 10.8, page 129

1 Think of any single significant mathematical development during that period which might be occupying the minds of mathematicians. You could look through the other units in this book. Does that give you any ideas?

Activity 10.9, page 130

1 Start by making a brief summary of the passage.

Activity 10.10, page 131

1 a Here is the translation of Arnauld's extract.

I do not understand why the square of -5 is the same as the square of $+5$, and that they are both 25. Neither do I know how to reconcile that with the basis of the multiplication of two numbers which requires that unity is to the one of the numbers what the other is to the product. This is both for whole numbers and for fractions. For 1 is to 3, what 4 is to 12. And 1 is to $\frac{1}{3}$ what $\frac{1}{4}$ is to $\frac{1}{12}$. But I cannot reconcile that with the multiplication of two negatives. For can one say that $+1$ is to -4, as -5 is to $+20$? I do not see it. For $+1$ is more than -4. And by contrast -5 is less than $+20$. Whereas in all other propositions if the first term is larger than the second, the third must be larger than the fourth.

Activity 10.11, page 133

1 a Summarise both extracts before you try to list the assumptions which each author makes. Here are notes on Euler's extract as an example:
- no doubt that positive multiplied by positive is positive, but separately examines positive by negative and negative by negative;
- negative numbers modelled as a debt; multiplying a debt by a positive number bigger than 1 gives a larger debt; thus $-a$ times b and a times $-b$ is $-ab$;
- $-a$ times $-b$ must be either $-ab$ or ab. It cannot be $-ab$ since $-a$ times b gives that and a times $-b$ cannot give same result; so $-a$ times $-b$ must be ab.

By looking through your summary can you see the assumptions which Euler and Saunderson make?

b You will first need to summarise Arnauld's objections to negative numbers and the assumptions which he makes. You can get help to do this if you consult the hint and also the answers to Activity 10.10.

To stage the debate, or to draw up a plan for your presentation or essay, you could consider the following format:
- first, summarise Arnauld's extract in Activity 10.10;
- secondly, summarise Euler's and Saunderson's extracts.

In each case you should draw out the key points. Then, you could start on the debate by
- presenting what Arnauld is likely to believe is the flaw in Euler's and Saunderson's arguments;
- allowing Euler and Saunderson to answer back or counterattack.

Remember not to use modern arguments to present the flaws in the various arguments. Try to argue from the points of view of Arnauld, Euler and Saunderson based on what you know of them from the extracts. You could consult your list of assumptions made by Euler and Saunderson; by looking at them can you find a 'chink in their armour' from Arnauld's point of view, or vice versa?

Activity 11.1, page 138

1 How did you set about establishing the result for the product of two negative numbers? You should consider whether the process you used was a justifiable one given the original role of the numbers a, b, c and d.

Activity 11.4, page 141

1 One counter-example is the answer given to question **1c** in Activity 10.10 in connection with Arnauld's opposition to negative numbers.

Activity 11.8, page 143

1 b The operations for combining the elements in Boole's algebra will not be labelled by + and × and the elements will not be labelled by numbers. Imagine replacing the symbols in the rules for combining Boole's elements by + and × in a suitable way and imagine the elements to be numbers. Would the rules then look the same as those for the number system? If not, how would they differ?

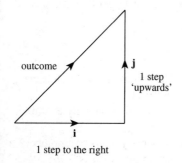

outcome

j

1 step 'upwards'

i

1 step to the right

Figure 13.1

Activity 11.10, page 145

1 To get started, here is part **b** as an example; you can then try the other parts.

Take a step of one unit to the right (**i**) and add to it a step of one unit 'upwards', (**j**). This produces a step as shown in Figure 13.1 but not in either of the basic directions (to the right and upwards); it is inclined at an angle to the basic directions.

What is the angle at which this new step is inclined and what is its length?

Activity 11.12, page 146

2 b Think about something you take for granted with real numbers. When you multiply three numbers together can you multiply them in any order; does the order matter?

14 Answers

Activity 2.1, page 13

1 a The table on the front tablet runs from 1, 2, …, 14 on the left and 9, 18, …, 126 on the right. The reverse side runs from 15 to 23 on the left and 135, in nines to 207 on the right.

b It is the nine times table.

2 The numbers from 1 to 23 are made by using the symbols ▼ and ◁ standing for 1 to 10. The number 63 appears to be made by using the symbol ▼ followed by a space, and then ▼▼▼. The number system therefore is based on base 60, with a space between the numbers in the 60s place and the units place.

3 698

4 The Old Sumerians used a system based on 10 and 60 using symbols, instead of place value, to represent larger numbers. The Babylonians adapted this to a system using only two symbols, and relative position to indicate what the symbols meant. However, there was still ambiguity in the Babylonians' system.

Activity 2.2, page 14

1 It is the 25 times table. The numbers to the left are 1 to 15, and those to the right are 25, 50, …, 375.

2 The right-hand number on the first line is ◁◁▓, while the right-hand side of the penultimate line is ▓◁▓. The last line is ◁▓ ▓◁.

3 The fifth number in the left-hand column is 5 while the twelfth number in the right-hand column is 300, which is 5×60. It is therefore represented in base 60 by the symbol for 5 followed by a blank space.

4 This is not easy to see, and you need to know from questions **1**, **2** and **3** what to look for. However, you can just make out the increasing numbers of the symbols ▼ and ◁ standing for 1 and 10 as you work downwards.

Activity 2.3, page 15

1 The left-hand column reads 2, 3, 4, 5, 6, 8, 9, 10, 12 while the right-hand column reads 30, 20, 15, 12, 10, 7 and 30, 6 and 40, 6, 5.

2 The product of corresponding numbers in the two columns is 60.

3 The sixth and seventh numbers in the right-hand column (far right) are the fractions $\frac{30}{60}$ and $\frac{40}{60}$.

4 It may be a table of lengths and breadths of rectangles whose areas are 60.

5 The two numbers 7 and 11 may have been omitted because fractions with denominators of 7 and 11 do not translate into finite sexagesimals.

Activity 2.4, page 15

1 The first could mean 2 and the second could mean 661.

2 ▼▼ ◁◁◁▓ and ▼▼ ◁◁◁▓.

3 It could mean 2, 61, 120 or 3600.

4 The columns of numbers are

2401	49
2500	50
2601	51
…	…
3364	58
3481	59
3600	60

The words will have something to do with square and square root.

Activity 2.5, page 17

1 7248.2, 3.155

2 2,46,40 and 1,0;36

3 $0;6,40$ and $0;2,13,20$

4 The only fractions in the decimal system which convert to terminating decimals are those with denominators which have prime factors consisting only of combinations of 2s and 5s; that is, the divisors of 10.

5 a Similarly, in the sexagesimal system, the fractions which terminate when converted to sexagesimals are those with denominators consisting of combinations of 2, 3, and 5, the prime factors of 60.
b More fractions have sexagesimals which terminate in the sexagesimal system.

6 $6\frac{5}{32} = 6;9,22,30$, $83\frac{1}{72} = 83;0,50$

7 $\frac{1}{7} = 0;8,34,17,8,34,17,\ldots.$ This lies between $0;8,34,18$ and $0;8,34,16,59$.

Activity 2.6, page 19

1 The addition sign is ⋈ and the multiplication sign is ⋈. The answers on the tablet are
a 3, 30, 15
b 15
c 3

2 See, for example, lines 9, 10 and 11, where it shows $27 - 15 = 12$.

3 a The reciprocal of 9 is $0; 6,40$. Then $47 \times 0; 6,40 = 5; 13,20$.
b $16; 4,46,52,30$

Activity 2.7, page 20

1 They are $0;30$ on the side of the square, and $1;24,51,10$ and $0;42,25,35$ underneath. The ambiguities associated with Babylonian numbers make it difficult to be certain what is intended, but you will see later that this is correct.

2 $0;30$ is the length of side of the square. The number above the diagonal is the length of the diagonal of a square of side 1. The number below, which is exactly one half of it, is the length of the diagonal of the given square, so the exercise carried out by the student is, given the length of the diagonal of a square of side 1, find the length of the diagonal of a square of side $0;30$, that is $\frac{1}{2}$.

3 The lower number is half the upper number; it is also its reciprocal.

4 ⊻

Activity 2.8, page 20

1 Let each side of the large square be 2 units. Then the area of the diagonal square is half the total area, so has area 2. The length of its side is therefore $\sqrt{2}$.

2 The figure shows an isosceles right-angled triangle outlined in black, with squares in grey drawn on its two adjacent sides. The diagonal square is also drawn. Pythagoras's theorem follows from the fact that the diagonal square is equal in area to the two grey squares.

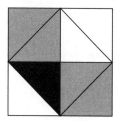 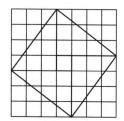

3 The right-hand figure shows that the area of the oblique square is $4 \times 6 + 1 = 25$ small squares, so the length of side of the oblique square is 5 units. But the squares on the short sides of the 3, 4, 5 triangle in the top left-hand corner of the figure are of area 9 and 16 which add to 25, thus verifying Pythagoras's theorem.

4 A search method which shows successively that $1 < \sqrt{2} < 2$, $1;24 < \sqrt{2} < 1;25$, $1;24,51 < \sqrt{2} < 1;24,52$ and $1;24,51,10 < \sqrt{2} < 1;24,51,11$ is a possibility.

Activity 2.9, page 21

1 a $1\frac{1}{2}$ cm by $1\frac{1}{3}$ cm, or $1;30$ cm and $1;20$ cm.
b Two more iterations are required. The approximation is not exact, but the expansion of $\frac{577}{408}$ gives $1;24,51,11$ correct to the nearest whole number, the true value of the last integer being $10\frac{36}{60}$.
c The successive approximations are $1;30$ and $1;25$.

2 The algorithm works quickly and effectively whatever the starting number.

3 This is the same algorithm as Newton's algorithm, $u_n = \frac{1}{2}\left(u_{n-1} + \frac{a^2}{u_{n-1}}\right)$ for calculating the square root of a^2.

4 The first approximation is $a + \frac{h}{2a}$, the second is
$$\frac{1}{2}\left(a + \frac{h}{2a} + \frac{a^2 + h}{a + \frac{h}{2a}}\right) = \frac{1}{2}\left(a + \frac{h}{2a} + \frac{2a(a^2 + h)}{2a^2 + h}\right).$$

Activity 2.10, page 22

1 $0; 30^2 + 0; 30 = 0; 15 + 0; 30 = 0; 45$
$30^2 - 30 = 15, 0 - 30 = 14, 30$

2 $x^2 + x = 0; 45$

3 The steps are
$$\left(x^2 + x + 0; 15\right) = 1$$
$$(x + 0; 30)^2 = 1$$
$$(x + 0; 30) = 1$$
$$x = 0; 30$$

4 To solve $x^2 - bx = c$, take the coefficient of x, that is b (ignoring the sign), halve it, $\frac{1}{2}b$, and square it to get $\frac{1}{4}b^2$. Add it to c to get $\frac{1}{4}b^2 + c$ which is the square of $\sqrt{\frac{1}{4}b^2 + c}$. Add the $\frac{1}{2}b$ to get $\frac{1}{2}b + \sqrt{\frac{1}{4}b^2 + c}$, which is the result.

5 It is the algorithm of al-Khwarizmi in the *Equations and inequalities* unit in Book 3.

6 The steps are
$$x^2 - bx + \left(\frac{1}{2}b\right)^2 = \frac{1}{4}b^2 + c$$
$$\left(x - \frac{1}{2}b\right)^2 = \frac{1}{4}b^2 + c$$
$$x - \frac{1}{2}b = \pm\sqrt{\frac{1}{4}b^2 + c}$$
$$x = \frac{1}{2}b \pm \sqrt{\frac{1}{4}b^2 + c}$$

7 The ancient method is very similar to the modern one, except that the idea of completing the square is kept hidden, and the whole appears very much like a recipe, rather like using the quadratic equation formula when you don't know where it has come from. The major difference is that the negative root is ignored.

8 $11x^2 + 7x = 6; 15$ so $(11x)^2 + 77x = 1, 8; 45$. Let $y = 11x$, so $y^2 + 7y = 1, 8; 45$. The coefficient is 7, so half is $3; 30$. $3; 30 \times 3; 30 = 12; 15$. Add $12; 15$ to $1, 8; 45$ to give $1, 21$. The square root is 9. As $9 - 3; 30 = 5; 30$, $y = 5; 30$ and $x = 0; 30$.

Activity 2.11, page 23

1 a $x^2 + ax = b$
b It is already in the list as type 1.

2 $\frac{1}{2}(x - y) = 3; 30$ so $\left(\frac{1}{2}(x - y)\right)^2 = 12; 15$, and $\left(\frac{1}{2}(x - y)\right)^2 + xy = 1, 12; 15$. Therefore

$\sqrt{\left(\frac{1}{2}(x - y)\right)^2 + xy} = 8; 30$ so
$\frac{1}{2}(x + y) - \frac{1}{2}(x - y) = 8; 30 - 3; 30$, so $x = 12$ and $y = 5$.

3 $\frac{1}{2}(x + y) = 3; 15$ so $\left(\frac{1}{2}(x + y)\right)^2 = 10; 33, 45$, and $\left(\frac{1}{2}(x + y)\right)^2 - xy = 3; 3, 45$. Therefore
$\sqrt{\left(\frac{1}{2}(x + y)\right)^2 - xy} = 1; 45$ so
$\frac{1}{2}(x + y) + \frac{1}{2}(x - y) = 3; 15 + 1; 45$ and
$\frac{1}{2}(x + y) - \frac{1}{2}(x - y) = 3; 15 - 1; 45$, so $x = 5$ and $y = 1\frac{1}{2}$.

Activity 2.12, page 25

1 The only alternative is to use the Greek construction methods which were not, so far as is known, available to the Babylonians.

2 The radius is $31; 15$ as on the tablet.

Activity 2.13, page 26

1

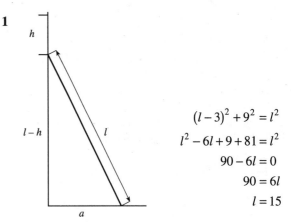

$$(l - 3)^2 + 9^2 = l^2$$
$$l^2 - 6l + 9 + 81 = l^2$$
$$90 - 6l = 0$$
$$90 = 6l$$
$$l = 15$$

Add 3^2 to 9^2 to find $6l$. Then divide by 2, then by 3 to find l.

2

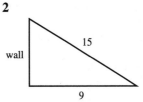

$15^2 = 3, 45$ and $9^2 = 1, 21$. Then $2, 24 = (\text{wall})^2$. To find wall, take the square root of $2,24$.

Activity 3.1, page 31

2 ABDC is a rhombus, and the diagonals of a rhombus cut at right angles.

3 If AD intersects the line l in E, the circle with centre A and radius AE is the circle required.

4 You are not allowed to draw this circle directly because, although you know its centre, you don't know its radius. Constructing the perpendicular to l through A enables you to find its radius.

Activity 3.2, page 31

1 a Draw two equal circles with centres A and B, intersecting at C and D. Then CD is the required perpendicular bisector.
b Both constructions rely on the geometric properties of a rhombus for their justification.

2 Draw a circle, centre A, to cut the lines at B and C. Then, with the same radius and centres B, C, draw two more arcs, cutting at D. Then AD bisects the angle at A.

3 The geometric properties of a rhombus explain all these constructions.

4 These constructions are left to you.

Activity 3.3, page 32

1 a A line through D is drawn parallel to CE. BC is produced to meet this line at F.
b Both regions consist of the quadrilateral ABCE and a triangle, in the first case CDE, and in the second CFE. These triangles have the same base, CE, and the same height, the distance between the parallel lines.
c Bisect AG so $GH = \frac{1}{2}GA$.
d It has the same height, and half the base.
e If $a : x = x : b$, then $\dfrac{a}{x} = \dfrac{x}{b}$, so $x^2 = ab$.
f Triangle ABD is similar to BCD, so $\dfrac{BD}{DA} = \dfrac{CD}{DB}$.
g Draw an arc of a circle, with centre H and radius HI. Extend GH to intersect this circle outside GH at N. Then GN is a line segment divided at H into lengths a and b. Find the mid-point of GN, and construct a semi-circle with this point as centre. To find the mean proportional, construct a perpendicular to GN at H. The point of intersection of the perpendicular with the semi-circle is K and HK is the mean proportional of lengths a and b, since triangle GKN is right-angled at K. Now construct the required square on HK.

Activity 3.4, page 34

1 Antiphon's solution was to find a series of regular polygons whose areas get progressively closer to the area of the circle.

Bryson's solution was to find the average of the inside and outside squares shown in Figure 3.10a.

2 Aristotle believes that the methods of Antiphon and Bryson are not scientific enough. He believes that Antiphon's method assumes that at some point the last polygon will coincide with the circle. Bryson's method uses the false assumption that two things which satisfy the same inequalities must be equal.

3 In neither case is a proper construction using Euclid's rules given. In Bryson's case, the areas are not proved equal. In Antiphon's case, the algorithm is not finite.

Activity 3.5, page 36

2 If AB is the given straight line, set the compasses to have radius AB. Then, with centres A and B, draw arcs of circles intersecting at C. Then triangle ABC is equilateral.

3 Proposition 1.9 is the same as Activity 3.2, question 2. Proposition 1.4 also relies on the properties of a rhombus, namely that the diagonals bisect each other.

4 Q. E. F. stands for Quod Erat Faceat. It means 'Which was to be made'.

Activity 3.6, page 37

1 It is Pythagoras's theorem.

2

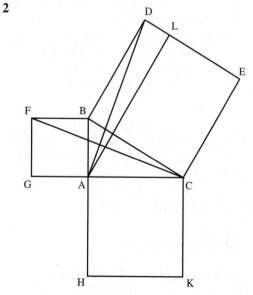

3 The proof is left to you.

Activity 3.7, page 38

1 A point is that which has no part. A line is breadthless length.

3 It is not clear what 'lying evenly' means.

4 A circle is a closed curve. It contains a point O inside it such that for all points X on the circle the lengths OX are equal to each other. O is called the centre of the circle. A diameter of the circle is any line through the centre which has its ends on the circle. The diameter also bisects the circle.

Activity 3.8, page 39

1 Postulate 3 says that if you are given the centre of a circle and its radius, you can draw the circle. Postulate 4 says that any right angle is equal to any other right angle.

2 If $a = b$, then $a + x = b + x$. If $a = b$, then $a - x = b - x$.

3 Postulate 5 gives a condition for two lines to meet, and where they should meet. But what happens if the 'angles are two right angles'?

Activity 4.1, page 42

1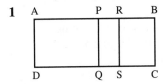

2 $AD \times (AP + PR + RB) =$
$$AD \times AP + AD \times PR + AD \times RB$$

3 $a(b + c + d) = ab + ac + ad$

4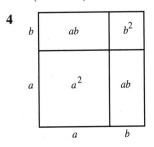

$$(a + b)^2 = a^2 + b^2 + 2ab$$

Activity 4.2, page 43

1

Euclid

Input A, B $\{A, B \text{ positive integers}\}$

```
repeat
    while A > B
        A − B → A
    endwhile
    if A = B
        then
            if B = 1
                then
                    print "A, B co-prime"
                else
                    print "A, B not co-prime"
            stop
        endif
    else
        A → T
        B → A
        T → B
    endif
until 0 ≠ 0
```

Output The highest common factor of A, B.

3 The 'greatest common measure' is the highest common factor.

4 In the algorithm in question **1**, replace the lines following **if** $A = B$ by
```
        then
            display B
        stop
```
and the output is the highest common factor of A and B.

Activity 4.3, page 43

1 Suppose that the number of prime numbers is finite, and can be written $a, b, \ldots, c$ where c is the largest. Now consider the number $f = ab \ldots c + 1$. Either f is prime or it is not.
Suppose first that f is prime. Then an additional prime f, bigger than c has been found.
Suppose now that f is not prime. Then it is divisible by a prime number g. But g is not any of the primes $a, b, \ldots, c$, for, if it is, it divides the product $ab \ldots c$, and also $ab \ldots c + 1$, and hence it must also divide 1, which is nonsense. Therefore g is not one of $a, b, \ldots, c$, but it is prime. So an additional prime has been found.
Therefore the primes can be written $a, b, \ldots, c, g$, which is more than were originally supposed.

Activity 4.4, page 45

1 Two lengths are commensurable if there is a unit such that each length is an exact multiple of that unit. If no such unit exists, the lengths are incommensurable.

2 Suppose that you can find a pair of positive integers a and b such that $a^2 = 3b^2$. Suppose also that a and b are the smallest such numbers. Then 3 divides a^2, so a is the square of a multiple of 3, so you can write $a = 3c$, so that $(3c)^2 = 3b^2$ or $9c^2 = 3b^2$. Cancelling the 3, you have $3c^2 = b^2$. But this pair of integers, b and c, is smaller than the original pair, which were supposed to be the smallest. This is a contradiction, so no such pair exists.

Activity 4.5, page 45

1 a *Heron*

Input A, x

> **repeat**
> > $x \to y$
> > $0.5(x + A/x) \to x$
>
> **until** $\mathrm{abs}(x - y) < 10^{-6}$

Output An approximation to $\sqrt{A}$.

b Starting with 1, $\sqrt{5} \approx 2.236\,067\,978$.

2 This method is the same as using Newton's method for finding the solution of the equation $x^2 - A = 0$, or the Babylonians' method for finding $\sqrt{A}$.

3 Change $0.5(x + A/x) \to x$ to $0.5(x + A/x^2) \to x$.

Activity 4.6, page 46

1 $S_0 = 3$ as the third side of the equilateral triangles is equal to the radius 1; triangles OAN and OXB are similar, and angles ONA and OBX are both right angles. Also OB bisects AC and XY. Use Pythagoras's theorem to find ON, and then use the similar triangles OAN and OXB to find $XB = \dfrac{1}{\sqrt{3}}$. This leads to the required result.

2 a The triangle BCD is an enlargement of the triangle formed by O, B and the foot of the perpendicular from O to the line BC. As the scale factor is 2, $CD = 2c_{n+1}$.
b The area of triangle BCD is either
$\frac{1}{2} \times BD \times NC = \frac{1}{2} \times 2 \times s_n = s_n$ or
$\frac{1}{2} \times BC \times CD = \frac{1}{2} \times 2s_{n+1} \times 2c_{n+1}$. The result follows.
c The result follows from $\dfrac{DC}{DN} = \dfrac{BD}{DC}$.
d The result follows from $\dfrac{BY}{BO} = \dfrac{NC}{NO}$.

3 a The lengths of each side of the inner and outer regular polygons are $2s_n$ and $2t_n$, and half the perimeter

means that there are $3 \times 2^{n+1}$ sides to be counted towards S_n and T_n.
b The first follows from questions **3a** and **2b** and the second from questions **3a** and **2d**.

4 *Archimedesπ*

Input None.

$3 \to S$	{Initialises S}
$6/\sqrt{3} \to T$	{Initialises T}
$\sqrt{3}/2 \to C$	{Initialises C}
$0 \to N$	{Initialises N, a counter}

> **repeat**
> > $\sqrt{(1+C)/2} \to C$ {Updates C}
> > $S/C \to S$ {Updates S}
> > $S/C \to T$ {Updates T}
> > $N + 1 \to N$ {Updates N}
> > display S
> > display T
> > display N
> > pause
>
> **until** $T - S < 10^{-10}$

Output The values of S and T which are both good approximations to π, and the number of iterations required to achieve them.

5 In modern trigonometric notation, you would probably write angle BOC as θ_n. The formulae from questions **2b**, **2c** and **2d** become

$$\sin \theta_n = 2 \sin \tfrac{1}{2}\theta_n \cos \tfrac{1}{2}\theta_n, \quad \cos \tfrac{1}{2}\theta_n = \sqrt{\frac{1 + \cos \theta_n}{2}} \text{ and}$$
$$\tan \theta_n = \frac{\sin \theta_n}{\cos \theta_n}.$$

Activity 4.7, page 48

1 a The y-value of Q is $\mathrm{f}\left(\tfrac{1}{2}(a+b)\right)$, and the y-value for M is the average of those for P and R, namely $\tfrac{1}{2}\left(\mathrm{f}(a) + \mathrm{f}(b)\right)$. Subtract these and the answer follows.
b The negative sign is there because when the parabola is drawn with its vertex upwards the coefficient of x^2 is negative, so p is negative. The expression $-\tfrac{1}{4}ph^2$ is therefore positive.
c From the expression $-\tfrac{1}{4}ph^2$ in part **a**, the length of QM in Figure 4.7 is independent of the precise values of a and b, but depends only on their difference h, that is the width of the step. If you halve the step, you replace h in the expression $-\tfrac{1}{4}ph^2$ by $\tfrac{1}{2}h$. Hence, the lengths UV and HK are both $-\tfrac{1}{4}p\left(\tfrac{1}{2}h\right)^2$, and are therefore equal.

2 a $y = px^2 + r$ is a parabola which has its vertex on the *y*-axis.

b The area of each of the triangles is $\frac{1}{2} \times a \times HK$, and $HK = \frac{1}{4}QM = \frac{1}{4}r$.

c There are two triangles of area $\frac{1}{8}ar$, four of area $\frac{1}{64}ar$, and so on.

d The sum of the GP is $\dfrac{ar}{1 - \frac{1}{4}} = \frac{4}{3}ar$, which is $\frac{2}{3}$ of the area of the rectangle which surrounds it.

Activity 4.8, page 50

1 The vertex of the cone is the fixed point. The axis is the straight line through the vertex and the centre of the defining circle.

2 See Figure 7.9 in this book.

Activity 4.9, page 50

1 a FH and MN are parallel, as are FG and AC. FH and LN are opposite sides of a parallelogram, and hence are equal in length.

b M, P and N are points on a semi-circle with diameter MLN, with PL perpendicular to MN. Angles MPN, MLP and PLN are all right angles, and triangles MPN, MLP and PLN are similar. Hence $\dfrac{ML}{LP} = \dfrac{PL}{LN}$, so $LP^2 = ML \times NL$.

c Triangles AHF and FLM are similar so $\dfrac{HF}{AH} = \dfrac{LM}{FL}$, and the result follows.

d Using the results of parts **a** and **c** in part **b**, $LP^2 = FL \times \dfrac{HF}{AH} \times FH$. Since the lengths AH and FH are independent of the position of MN, the value of $\dfrac{LP^2}{FL}$ is a constant, say *k*.

2 a The ratio $\dfrac{LM}{LF}$ is a constant as MN moves, and so is the ratio $\dfrac{LN}{LJ}$.

b Triangles FKJ and LQJ are similar, so $\dfrac{FK}{FJ}$ is equal to the constant of proportionality; so too is $\dfrac{QL}{LJ}$. Then $LP^2 = k \times LF \times LJ = QL \times LF$, and the square of LP is equal to the rectangle FLQR.

3

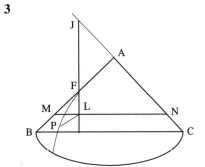

$LP^2 = LM \times LN$ is proportional to $LF \times LJ$, by the same argument as in question **2a**. Now suppose that the constant of proportionality is the ratio of some length FK to the length FJ as in the diagram below.

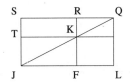

Then $LP^2 = \dfrac{FK}{FJ} \times LF \times LJ$, and there exists a point P such that $\dfrac{FK}{FJ} = \dfrac{LQ}{LJ}$. So $LP^2 = \dfrac{LQ}{LJ} \times LF \times LJ = LQ \times LF$. Hence 'the square of LP is applied to the line FJ in hyperbola (in excess or thrown beyond) by a rectangle similar to the given rectangle JFKT'.

Activity 5.1, page 57

1 The widow's share is $\frac{1}{8}$ of the estate. The sons inherit $\frac{1}{4}$ each of the remainder, that is, $\frac{7}{32}$ parts. So the stranger gets $\frac{7}{32} - \frac{1}{8} = \frac{3}{32}$, leaving $\frac{29}{32}$ for the family. The wife gets $\frac{1}{8} \times \frac{29}{32} = \frac{29}{256}$. The sons get $\frac{7}{32} \times \frac{29}{32} = \frac{203}{1024}$. So the estate is divided into 1024 equal parts. The stranger gets 96, the wife 116, and the sons 203 each.

2 The stranger, sons and daughters receive their shares in the ratio $2 : 2 : 2 : 1$ so the estate is divided into 7 equal parts; the daughter gets one part, and the rest 2 parts.

3 The mother's share is $\frac{1}{6}$. The children's shares would be $2 : 2 : 2 : 1 : 1$ so $\frac{5}{6}$ is shared in 8 parts. The stranger gets $\frac{5}{48}$ of the estate. There was no second daughter, so the children share the residue in the ratio $2 : 2 : 2 : 1$. The estate is divided into 2016 parts. The stranger gets 210, the mother 301, each son gets 430 and the daughter 215.

14 Answers

4 If he had a daughter the shares would be $2:2:2:1$ so the stranger gets $\left(\frac{2}{7}-\frac{1}{7}\right)+\frac{1}{3}\left(\frac{1}{3}-\frac{1}{7}\right)$ or $\frac{13}{63}$, leaving $\frac{50}{63}$ to be shared by his three sons. The estate is divided into 189 equal parts: the stranger gets 39 and the sons 50 each.

Activity 5.2, page 58

1 **a** crd $36° = 2\sin 18° = 0.618\,033\,989$. Converting this to sexagesimal, you find that it is $0;37,4,55$, the difference between the two values being 1.58×10^{-6}.
b The next sexagesimal figure is 20.

2 Use the expression $2x = $ crd $(2(90-50))°$.

3 What you need is $\cos 35°$ so that you can use the cosine formula. Use the fact that $\cos 35° = \sin 55° = \frac{1}{2}$ crd $110°$.

Activity 5.3, page 58

1 $\sin 30° = \frac{1}{2}$, $\cos 30° = \frac{1}{2}\sqrt{3}$, $\sin 45° = \cos 45° = \frac{1}{2}\sqrt{2}$, $\sin 60° = \frac{1}{2}\sqrt{3}$, $\cos 60° = \frac{1}{2}$.

2 **a** Using isosceles triangles, BC = BD = AD = 1. Then $AB = 2\cos 36°$ and $AC = 1 + 2\cos 72°$.
b Writing $c = \cos 36°$, you get the quadratic equation $4c^2 - 2c - 1 = 0$, and you need the positive solution.
c Using the relation $\cos 2\alpha = 2\cos^2 \alpha - 1$ again, you find that $2\cos^2 18° - 1 = \frac{1}{4}\left(1+\sqrt{5}\right)$, from which the result follows.
d Use $\cos^2 18° + \sin^2 18° = 1$.

3 There are a number of possibilities. They actually used the fact that $\sin 72° = \cos 18°$ and $\cos 72° = \sin 18°$ together with the values of $\sin 60°$ and $\cos 60°$ in the formula for $\sin(72° - 60°)$.

4 $0.017\,452\,05$. This is correct to 6 decimal places.

5 **a** $0.017\,452\,406$. The ninth place doesn't change after four iterations.
b The convergence is by no means as quick. You need about 12 iterations to achieve an accurate result.
c Convergence is very slow indeed.

Activity 5.4, page 60

1 He probably summed the GP $1+2+4+\ldots+2^{63} = 2^{64} - 1$ and then calculated 2^{64} by repeated squaring.

2 **a** $(14,2)$, $(10,5)$, $(6,8)$ and $(2,11)$

b Because the solutions cannot be determined precisely.

3 The algorithm is given in the hints section. There are 26 solutions in all.

4 **a** 204, 577 and 1189, 3363
b The method does apply to $2x^2 + 1 = y^2$. Successive solutions are 2, 3; 12, 17; 70, 99; 408, 577. However, the method doesn't help to find a starting solution.

Activity 5.5, page 61

1 The factors of 220 are 1, 2, 4, 5, 10, 11, 20, 22, 44, 55, 110 which add to 284. The factors of 284 are 1, 2, 4, 71, 142 which add to 220.

2 **a** When $n = 2$, $p = 5$, $q = 11$ and $r = 71$ which are all prime. The values of M and N are then 220 and 284. When $n = 3$, you obtain $M = 2024$, $N = 2296$, and when $n = 4$, $M = 17\,296$ and $N = 18\,416$.
b Provided p and q are prime, the factors of M which are less than M are 1, 2, 2^2, …, 2^n, p, $2p$, $2^2 p$, …, $2^n p$, q, $2q$, $2^2 q$, …, $2^n q$, pq, $2pq$, $2^2 pq$, …, $2^{n-1}pq$. There are four geometric progressions here which sum to N. Showing that the factors of N sum to M is easier.

Activity 5.6, page 62

1 The first approximation is to see that the square root is between 200 and 300. Substituting $y = x - 200$ gives $y^2 + 400y = 31\,824$ with a solution between 0 and 100. Inspection shows that it is between 60 and 70, so write $z = y - 60$ giving $z^2 + 520z = 4224$ with a solution between 0 and 10. Inspection now shows that 8 fits exactly, showing that $\sqrt{71\,824} = 268$.

2 **a** The first approximation is 30 and the second is 7. At this stage the equation is $z^2 + 51z = 42$, and the solution is approximately $z = \frac{42}{51+1} = 0.81$. This leads to $x \approx 37.81$.
b The first approximation is 20, and 4 is a solution of the remaining equation. The solution is 24.

Activity 5.7, page 63

1 Suppose that the shortage and surplus are e_1 and e_2, so that they are both positive. Then the algorithm says that $x = \frac{x_1 e_2 + x_2 e_1}{e_1 + e_2}$.

2 The result follows from the equation $\frac{x - x_1}{e_1} = \frac{x_2 - x}{e_2}$ arising from similar triangles.

3 a You get the answer 2 tou and 5 sheng.

b 1.32

Activity 5.8, page 65

1 $x^2 + 10x = 39$ so $x^2 + 10x + 25 = 64$, leading to $(x + 5)^2 = 64$. Remembering that only positive numbers were allowed at the time, $x + 5 = 8$, so $x = 3$.

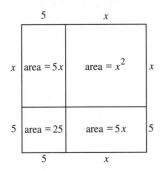

The total area is $(x^2 + 10x) + 25 = 39 + 25 = 64$ so the side of the largest square is 8, leading to $x = 3$.

3 When you express the rule in symbols you obtain $x(n - x) + \left(\frac{1}{2}n - x\right)^2 = \left(\frac{1}{2}n\right)^2$. When you multiply out the brackets, all the terms cancel.

4 No answer is given to this question.

5 Abu Kamil uses a rigorous Euclidean approach which is difficult to follow geometrically. Al-Khwarizmi's solution is clearer geometrically, and less rigorous.

Activity 5.9, page 68

1 a By drawing the graphs of $y = x^2$ and $y = \dfrac{20 - x}{x}$, the point of intersection is at $x = 2.59$.

b By drawing the graphs of $y = x^2$ and $y = \dfrac{20}{x - 1}$ you find $x = 3.09$.

Activity 5.10, page 69

1 The function takes its maximum value in the range $0 < x < a$ when $x = \frac{2}{3}a$.

2 Al-Tusi may have known that the arithmetic mean (AM) of positive numbers is always greater than or equal to the geometric mean (GM), that is, that

$$\frac{a_1 + a_2 + \ldots + a_n}{n} \geq \left(a_1 a_2 \ldots a_n\right)^{\frac{1}{n}}.$$ Now choose the three numbers $a - x$, $\frac{1}{2}x$ and $\frac{1}{2}x$. Their AM is $\frac{1}{3}a$, and their

GM is $\left(\frac{1}{4}x^2(a - x)\right)^{\frac{1}{3}}$. Then $\frac{1}{3}a \geq \left(\frac{1}{4}x^2(a - x)\right)^{\frac{1}{3}}$ leads to the conclusion.

3 Look at the graphs of $y = x^2(a - x)$ and $y = c$ and see where they intersect.

Activity 6.2, page 74

1 Use compasses with a radius greater than $\frac{1}{2}AB$ and draw arcs with centres A and B, intersecting at points C and D. Then join CD.

2 Suppose that X is any point on CD.

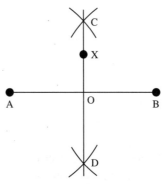

Then $AX^2 = AO^2 + OX^2$
$$= AC^2 - CO^2 + OX^2$$
$$= BC^2 - CO^2 + OX^2$$
$$= BO^2 + OX^2$$
$$= BX^2$$

Suppose now that X does not lie on CD and that the foot of the perpendicular from X to AB is N so that $AN \neq BN$. Then $AX^2 = AN^2 + NX^2$
$$\neq BN^2 + NX^2$$
$$\neq BX^2$$

The combination of these two arguments shows that only points on CD satisfy the condition.

3 The answer is in the text following Activity 6.2.

4 The answer is in the text following Activity 6.2.

Activity 6.4, page 77

1 $a = 2$

2 $a = 3.46$

3 The positive solution of $a^3 + 6a^2 + 12a = 90$ which is about $a = 2.61$.

14 Answers

Activity 6.5, page 79

1 a $A^2 + 2BA = Z$ or $x^2 + 2bx = c$

b $2DA - A^2 = Z$ or $2dx - x^2 = c$

c $A^2 - 2BA = Z$ or $x^2 - 2bx = c$

d Viète takes Z to represent a plane figure.

e Viète still doesn't allow negative coefficients or unknowns.

Activity 6.6, page 80

1 The lengths of certain straight lines are the basic variables.

Activity 6.7, page 81

1 $\sqrt{a}$

2 a $x = ab$

b $x = \dfrac{b}{a}$

c $x = \sqrt{a}$

d $x = \sqrt[3]{a}$ or $x = a^{\frac{1}{3}}$, $y = \left(\sqrt[3]{a}\right)^2$ or $y = a^{\frac{2}{3}}$

3 a '… having given two other lines, to find a fourth line which shall be to one of the given lines as the other is to unity (which is the same as multiplication)'

b '… to find a fourth line which is to one line as unity is to the other (which is equivalent to division)'

c, d '… to find one, two, or several mean proportionals between unity and some other line (which is the same as extracting the square root, cube root, etc., of the given line)'

Activity 6.8, page 81

1 From similarity, $\dfrac{AB}{BC} = \dfrac{DB}{BE}$. As $AB = 1$, $BC = \dfrac{BE}{DB}$.

2 The triangles FIG and IHG are similar, so $\dfrac{GI}{FG} = \dfrac{GH}{GI}$. As $FG = 1$, $GI = \sqrt{GH}$.

Activity 6.9, page 82

1 It is sufficient to designate each line by a different single letter.

2 'Here it must be observed that by a^2, b^3, and similar expressions, I ordinarily mean only simple lines, which, however, I name squares, cubes etc., so that I may make use of the terms employed in algebra.'

3 '… to find one, two, or several mean proportionals between unity and some other line …' from paragraph 2, and 'If the square root of GH is desired, …' from paragraph 3.

Activity 6.10, page 82

1 Descartes multiplies or divides each term in the equation by unity an appropriate number of times. For example, Descartes considers $\sqrt[3]{a^2 b^2 - b}$, and says that $a^2 b^2$ is divided once by unity and b is multiplied twice.

Activity 6.11, page 84

1 The sketch of the solution illustrates the first stage. Descartes gives letters to lines. (This is the second stage.) The third stage is the derivation of the equation

$$\sqrt{x^2 + y^2} = \sqrt{(d-x)^2 + y^2}.$$

Manipulating to get the equation $x = \frac{1}{2}d$ is the fourth stage.

The final stage is 'You can see that the solution in this case is the perpendicular on AB at a distance equal to $\frac{1}{2}d$ from A'.

2 The idea of using equations, and sets of equations, for calculating geometric quantities, though commonplace now, was quite new to Descartes.

3 Known quantities are designated by the first letters of the alphabet, and unknown quantities by the last letters.

Activity 6.12, page 84

1 $x(a + b + x) = (b + x)^2$

2 $x = \dfrac{b^2}{a - b}$

Activity 6.13, page 85

1 Angle QNR stands on the chord PQ, so the angle at the centre also on PQ, that is angle POQ, is double angle QNR. The fact that the triangles NOQ, QOT and TOP are congruent shows that angles NOQ, QOT and TOP are all equal. The result follows.

2 Triangles NOQ and QNR are similar because two angles are equal. (Angle NOQ = angle QNR and angle NQO is common to both triangles.)

Angle ORM = angle NRO, and angle ROM = angle QNR so triangle ROM is similar to triangle QNR.

3 As NQ = NR, since triangle QNR is similar to the isosceles triangle NOQ, RM = $q - 2z$.

---

4 From triangles NOQ and QNR, $\dfrac{QR}{z} = \dfrac{z}{1}$, so $QR = z^2$.

From triangles ORM and QNR, $\dfrac{RM}{QR} = \dfrac{OR}{NQ}$, so

$\dfrac{q - 2z}{z^2} = \dfrac{1 - z^2}{z}$ from which it follows that $z^3 = 3z - q$.

Activity 7.1, page 88

1 If $z^2 = ab$, then z is the side of a square which is equal in area to a rectangle with sides a and b.

Take $a + b$ as the diameter of a circle. P is the point where the lines of length a and b join. Draw PQ perpendicular to the diameter. Then $PQ = z$.

2 Take an arbitrary acute angle with the lengths a and b on the arms. Mark off the length b on the arm with the length a leaving a length $a - b$ as shown in the figure.

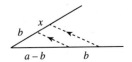

Now draw two similar triangles, and x is the required length.

Activity 7.2, page 89

1 Lengths which are solutions of quadratic equations.

2 Paragraph 13 makes it clear that, at the very least, it is hard to construct the solutions of cubic equations with a straight edge and compasses. In fact, it is impossible.

Activity 7.3, page 89

1 In triangle LMN, $MN^2 - NL^2 = ML^2$. Therefore $(MN - NL)(MN + NL) = ML^2$, or, since $NL = NP$, $MP.MO = ML^2$

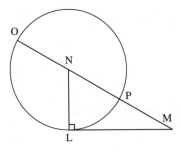

2 Put $LM = b$ and $LN = \frac{1}{2}a$. Then $OM = z$ because, with $MP.MO = ML^2$, $z(z - a) = b^2$ or $z^2 = az + b^2$.

3 The second solution is negative, and therefore was not considered by Descartes as a possible length of a line segment. It can only be acceptable when you consider directed line segments, or vectors, and take into account the idea that the direction provides the sign of the number. If you do allow negative lengths, then you can check that the other solution corresponds to MP.

Activity 7.4, page 90

1 This is very like Activity 7.3.

In the diagram for Activity 7.3 (see above), put $MP = z$. Then $MP.MO = ML^2$ gives $z(z + a) = b^2$. Then use the result of question **1** in Activity 7.1 to construct the solution of $x^2 = z = z \times 1$, where z is known.

2 The equation $x^4 = -ax^2 + b^2$ does not satisfy the law of homogeneity, unless you consider x^2 as being divided once by unity.

Activity 7.5, page 90

1 Triangle NQR is isosceles, since $PN = QN$. Since O is the mid-point of QR, N is on the perpendicular bisector of QR. And because LM is perpendicular to LN, it is also perpendicular to MR or QR. Since NO and LM are both perpendicular to QR, they are parallel.

2 $OQ = \sqrt{NQ^2 - NO^2} = \sqrt{\frac{1}{4}a^2 - b^2}$

3 $MQ = MO - OQ = \frac{1}{2}a - \sqrt{\frac{1}{4}a^2 - b^2}$

$MR = MO + OQ = \frac{1}{2}a + \sqrt{\frac{1}{4}a^2 - b^2}$

4 Because both solutions are positive.

5 In the case when the circle doesn't intersect or touch the line MR; that is, when $b > \frac{1}{2}a$. Descartes describes this in paragraph 17.

6 The equation $z^2 + az + b^2 = 0$ has solutions $z = -\frac{1}{2}a \pm \sqrt{\frac{1}{4}a^2 - b^2}$. Both these solutions are always negative.

Activity 7.6, page 91

1 A method for the solution of all construction problems, which is generally applicable.

Activity 7.7, page 92

1 Choose values yourself for the number of unknown quantities, to which no equation corresponds.

Activity 7.8, page 92

1 The coordinates of X are (x, y). As distance $AX = 2 \times$ distance BX, you have $\sqrt{x^2 + y^2} = 2\sqrt{(x-a)^2 + y^2}$ which leads to $x^2 + y^2 = 4\left((x-a)^2 + y^2\right)$ or $3x^2 + 3y^2 - 8ax + 4a^2 = 0$.

2 Write the equation as $x^2 + y^2 - \frac{8}{3}ax + \frac{4}{3}a^2 = 0$ or as $\left(x - \frac{4}{3}a\right)^2 + y^2 = \frac{4}{9}a^2$. This shows that the curve is a circle, centre $\left(\frac{4}{3}a, 0\right)$, radius $\frac{2}{3}a$.

Activity 7.9, page 92

1 Let $PA = x$ and $PC = y$. This is, in effect, taking n as the x-axis and l as the y-axis. Then, if $PA^2 = PB \times PC$, then $x^2 = |x - a| \times |y|$, the modulus signs showing that the positive value of the distance is always taken. Then $y^2 = \dfrac{x^4}{(x-a)^2}$ or $y = \dfrac{x^2}{|x-a|}$ or $y = -\dfrac{x^2}{|x-a|}$.

2 Take $a = 1$. Then the graph is a hyperbola and its reflection in the x-axis.

Activity 7.10, page 95

1 Descartes wants to go further, and to distinguish several other classes of problem within the linear class.

2 In antiquity, the linear curves were defined as mechanical because they could only be drawn with an instrument. In Descartes's opinion this is remarkable because the straight line and the circle need instruments, the straight edge and compasses. The fact that instruments are more complicated, and the resulting curves therefore more inaccurate is, according to Descartes, no reason to call the curve mechanical.

3 Between any pair of points a straight line always can be drawn.
Through any point a circle can be drawn with a given centre.

4 Descartes says that the mathematicians of antiquity did not restrict themselves to these two postulates. For instance, they assumed that any given cone can be intersected by a given plane. He uses the argument here because he wants to plant an idea which he will use to back his later suggestion to accept construction curves other than the line and circle.

5 The curve must be described by a continuous motion or by a succession of several continuous motions, each motion completely determined by those which precede it.

Activity 7.11, page 95

1 Q and R are the feet of the perpendiculars from P to the y-axis and the x-axis.

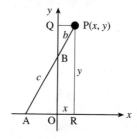

As triangle PAR is similar to triangle BAO, $\dfrac{RP}{OB} = \dfrac{PA}{BA}$, so $y = \dfrac{a}{c}y_B$. Similarly, it follows from the similar triangles PBQ and ABO that $x = -\dfrac{b}{c}x_A$.

2 $OA^2 + OB^2 = AB^2$ so $\dfrac{c^2}{b^2}x^2 + \dfrac{c^2}{a^2}y^2 = c^2$. Therefore the coordinates (x, y) of P satisfy the equation $\dfrac{x^2}{b^2} + \dfrac{y^2}{a^2} = 1$. This is the equation of an ellipse.

3 Yes. When the pin A, for example, moves along the *x*-axis, the straight line motion of pin B is directly linked to it. The curve traced out by P is directly linked to these two preceding motions.

Activity 7.12, page 96

1 Take, for instance, EG as the positive *x*-axis and EC as the positive *y*-axis. FD intersects BG in R, as shown in the figure.

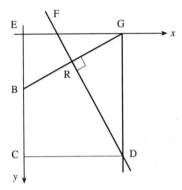

Suppose that EG = x, EC = y and EB = p.

BG is perpendicular to FD, and triangle BEG is similar to triangle GRD. From this you can deduce that $\frac{1}{2}(p^2 + x^2) = py$. This is the equation of a parabola because you can write it in the form $x^2 = 2p(y - \frac{1}{2}p)$.

3 This instrument satisfies the conditions. The motion of FK is directly linked to the straight line motion of G, and the motion of D is in turn directly linked to the motion of the ruler FK, and of GP.

Activity 7.13, page 98

1 Triangles BYC and CYD are similar so $\dfrac{x}{y} = \dfrac{a}{BC}$, and from triangle BYC, $BC^2 = x^2 - a^2$.

Activity 7.14, page 99

1 He forgot the straight line.

Activity 7.15, page 99

1 $\left(x + \frac{1}{2}q\right)^2 + (y + 2)^2 = \left(\frac{1}{2}q\right)^2 + 4$ or $x^2 + y^2 + qx + 4y = 0$

2 The *x*-values of g, G and F are the points of intersection of $x^2 + y^2 + qx + 4y = 0$ and $y = -x^2$. Therefore, substituting $y = -x^2$, you find $x^2 + x^4 + qx - 4x^2 = 0$, that is, $x^4 - 3x^2 + qx = 0$ or

$x(x^3 - 3x + q) = 0$. As $x = 0$ is not a suitable solution, you are left with $x^3 - 3x + q = 0$.

Activity 8.2, page 102

1 The degree of the equation of the curve defines the class of the curve.

Activity 8.3, page 103

1 'When the relation between all points of a curve and all points of a straight line is known, in the way I have already explained, it is easy to find the relation between the points of the curve and all other given points and lines.'

Activity 8.4, page 103

1 To find the angle formed by two intersecting curves.

Activity 8.5, page 104

1 $x^3 - 3x^2 - 2x + 6 = 0$

Activity 8.6, page 105

1 He calls these negative solutions 'false roots'.

2 The defect of a quantity 5, is the number –5.

Activity 8.7, page 105

1 One positive solution; two positive solutions

2 See the remarks immediately following the activity.

Activity 8.8, page 106

1 The equation has two positive solutions. The sign rule works.

2 Two changes of sign, so two positive solutions.

3 The *x*-coordinate of F.

Activity 8.9, page 108

1 The equation can be written in the form $\dfrac{(x-2)^2}{2^2} + \dfrac{y^2}{1^2} = 1$.

2 a The circle has centre $(p, 0)$ and radius $\sqrt{(3-p)^2 + \frac{3}{4}}$. Its equation is therefore $(x-p)^2 + y^2 = (3-p)^2 + \frac{3}{4}$ which simplifies to $y^2 = -x^2 + 2px + 9\frac{3}{4} - 6p$.

b This meets the ellipse where
$\frac{1}{4}\left(4x - x^2\right) = -x^2 + 2px + 9\frac{3}{4} - 6p$ which simplifies to
$x^2 + \frac{1}{3}(4 - 8p)x + 8p - 13 = 0$.

c If 3 is a double solution, then this equation must be the same equation as $(x - 3)^2 = 0$. It follows that $\frac{1}{3}(4 - 8p) = -6$, and that $8p - 13 = 9$. Both equations give $p = 2\frac{3}{4}$.

3 The points $\left(2\frac{3}{4}, 0\right)$ and $\left(3, \frac{1}{2}\sqrt{3}\right)$ lie on the normal. The equation is therefore $y = 2\sqrt{3}x - \frac{11}{2}\sqrt{3}$.

4 P is the point $(2, p)$ on the axis of the parabola.

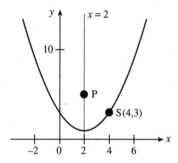

The circle with centre P and radius PS has equation
$(x - 2)^2 + (y - p)^2 = 4 + (p - 3)^2$. Eliminating $(x - 2)^2$ between the two equations you find
$y^2 + (2 - 2p)y + (6p - 15) = 0$. As this equation has the double solution $y = 3$, $2 - 2p = -6$ and $6p - 15 = 9$. Therefore $p = 4$, and the equation of the normal is $x + 2y = 10$.

Activity 8.10, page 108

1 In *La Géométrie*, one geometric property is being examined, the length of a line segment. This length is considered as the unknown in geometric construction problems.
Descartes proposed to represent a curve by means of an equation. He needed this to classify curves, and finally to find the most suitable curves for a construction.

2 The intention of *La Géométrie* is not to investigate the properties of curves, but to solve geometric construction problems. The curves are merely tools for solving these problems. A clear exception to this is his method of finding normals. In this case, he studies the direction of the normal at a point on a curve, for example an ellipse. In *La Géométrie*, x-y axes are not used, although in a number of cases the unknowns are named in such a way

that the result makes it appear as though a grid has been used.

Activity 8.11, page 109

1 On account of the law of homogeneity, the right-hand side of the equation must be quadratic. The addition 'pl' indicates that Z is not a linear but a plane magnitude (planus = plane).

2 a In this example, Fermat uses a kind of coordinate system, with NP and PI. The part of the curve that corresponds to negative values of A and E is not considered. The curve is limited to the first quadrant. However, Fermat also draws curves outside the first quadrant. For example, he continues to draw a complete circle, which only lies one quarter in the first quadrant, in spite of ignoring negative numbers.

b Compared with Descartes's mathematics:
- Fermat's mathematics obeys the law of homogeneity, whereas Descartes dropped it.
- Descartes ventures a little into using negative numbers, for example, as solutions to equations, but Fermat does not.
- Fermat and Descartes use different names for known magnitudes and constants.
- Descartes doesn't really investigate the graphs of curves, but Fermat does.

Compared with today's mathematics, Descartes approaches present-day mathematics in the case of the first three points, but in the fourth case, Fermat is more modern.

Activity 9.1, page 113

1 OP, angle OPG, OPG, two

Activity 9.2, page 113

1 a Let t stand for the time in seconds. Then, after t seconds the particle is at P in the diagram.

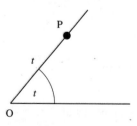

The coordinates of P are $(t\cos t, t\sin t)$, which leads to the result.

b $r = \theta$

2 $x = vt \cos \omega t$, $y = vt \sin \omega t$; $r = \dfrac{v\theta}{\omega}$

3 a $\left(v\omega t^2 \cos\left(\omega t - \tfrac{1}{2}\pi\right), v\omega t^2 \sin\left(\omega t - \tfrac{1}{2}\pi\right) \right)$ or $\left(v\omega t^2 \sin \omega t, -v\omega t^2 \cos \omega t \right)$

b Gradient PT $= \dfrac{-\omega t \cos \omega t - \sin \omega t}{\omega t \sin \omega t - \cos \omega t}$; from the

Cartesian parametric equation the gradient is $\dfrac{dy}{dt} \Big/ \dfrac{dx}{dt}$, which leads to the same result.

4 From the similar right-angled triangles $\dfrac{OT}{r} = \dfrac{r\omega}{v}$.
Using $r = vt$ leads to the result. The length of the arc PK is also $r\omega t$.

Activity 9.3, page 115

1 a $b(a+e) - (a+e)^2$

b $z - z' = e^2 + e(2b - x)$

c If the coefficient, $(2b - x)$, of e was negative, then you could find a small positive value of e such that $z < z'$, and z is not the maximum. Similarly, when the coefficient of e is positive. Therefore the coefficient of e is zero.

2 a $z' = a^2(X + e) - (X + e)^3$

b $z - z' = e^2 + e(3X^2 - a^2) + 3Xe^2 + e^3$

c $X^2 = \tfrac{1}{3}a^2$, giving the two solutions $X = \pm\tfrac{1}{\sqrt{3}}a$. You need to check which gives the maximum by another method.

Activity 9.4, page 116

1 a $NQ = \dfrac{(s+e)y}{s}$

b $x + e, \left\{ a(x+e)^2 \right\}^{\frac{1}{3}}$

2 $\dfrac{a}{s^3}\left\{ xs^2 e(3x - 2s) + se^2(3x^2 - s^2) + x^2 e^3 \right\}$

3 a $s = \tfrac{3}{2}x$

c Differentiate to find that $\dfrac{dy}{dx} = \dfrac{2ax}{3y^2}$, and then use the

equation of the curve to show that you can write this as $\dfrac{y}{s}$.

Activity 9.5, page 116

1 a $\left(x + e, \dfrac{(s+e)y}{s} \right)$

b $\left\{ \dfrac{(s+e)y}{s} \right\}^3 = a(x+e)^2$

c $e(3x^2 s^2 - 2xs^3) + e^2(3x^2 s - s^3) + x^2 e^3 = 0$

d $(3x^2 s^2 - 2xs^3) + e(3x^2 s - s^3) + x^2 e^2 = 0$

2 $s = \tfrac{3}{2}x$

Activity 9.7, page 118

1 w, $\overline{\text{omn. } w}$, $\overline{xw}$, ult. $x \times \overline{\text{omn. } w}$, $\overline{\text{omn.} \overline{\text{omn. } w}}$

2 $\text{OCD} = \sum xw$, $\text{ODCB} = x_{\max} \times \sum w$,
$\text{OCB} = \sum\sum w$ so $\sum xw = x_{\max} \times \sum w - \sum\sum w$.

Activity 9.8, page 119

1 $\dfrac{dy}{dx} = \dfrac{y}{t}$

2

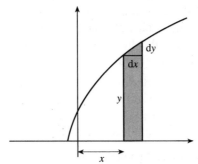

Activity 10.1, page 124

1 In a practical way negative numbers arose naturally in accounting and financial systems. This is the way that they were first used by the Chinese about 2000 years ago. A positive number might have been used to represent money received while a negative number represented the opposite, that is money spent.

2 a A consistent interpretation of addition is as a movement to the right of the number line through a certain distance.

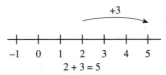

Thus you can justify that the result of adding 3 to 2 gives 5.

This model gives you the same result if you started with 3 and added 2 to it. Your model is therefore consistent with the rules of addition as it should be.

To get an interpretation of subtraction remember that the operation of subtraction is the inverse of addition. If you want to reverse, or undo, the result of adding a number then you subtract it again. If you model addition as a movement to the right on the number line, then it is natural to model subtraction as a movement to the left.

Suppose that you take 2 and subtract 3 from it. You start at the point corresponding to 2 and move through a distance of 3 units to the left, ending up at the point corresponding to -1. Thus your picture accurately models the arithmetic operation: $2 - 3 = -1$.

b You can derive an interpretation of multiplication of two positive numbers from your model of addition. Multiplying a positive number a by another positive number b is equivalent to adding the number b to the number zero a times. In this picture, to consider the result of 2×3, for example, start at the point 0 and move through a distance of 3 to the right, ending up at the interim point 3. Then move from the interim point 3 through another 3 units to the right thus producing the final result of 6. On the other hand 3×2 involves starting at 0 and adding 2 in each of three stages.

This model can be extended to consider examples such as $2 \times (-3)$. Starting at 0, you add -3, or equivalently subtract 3, in each of two stages. This involves two movements to the left and produces the final point -6.

In this model it is difficult to come up with an interpretation of an operation such as $(-3) \times 2$. To be consistent with the picture used so far, you would have to start at 0 and then add 2 a certain number of times, but how many times corresponds to -3? Your model breaks down at this point and it can only come up with an answer for $(-3) \times 2$, if you assume that it has the same value as $2 \times (-3)$. But this assumption is based on things you know about multiplication from another source; it is not derived from your model. If you only draw on the properties of the model itself then the model fails at a particular point.

It is also difficult in this model to give an interpretation of the result of multiplying two negative numbers.

Other models of operations on the number line do permit a consistent interpretation of addition, subtraction, multiplication and division applied to all combinations of positive and negative numbers. You will meet one in Chapter 11. Until then you should bear in mind that some models of numbers only allow a limited interpretation of operations on the numbers.

Activity 10.2, page 125

1 You can use subtracted numbers to model practical problems since they have many practical applications; for example, the difference in the length of two lines, which arose quite naturally in geometric arguments proposed by the Greeks, and the amount of cereal remaining in the store after some had been removed during the winter to feed the people in a village.

You can see from these examples that the process of subtracting one positive number from another is much more acceptable. The practical application of the process means that it is not possible to subtract a larger number from a smaller. You should think of this in the context of the cereals. If the store contained 2 tons of rice and you needed 3 and you started to remove 3, then the store would be empty long before you had finished. Thus the subtracted number $2 - 3$ would not have had any practical solution. Using subtracted numbers does not imply the need for negative numbers; subtracted numbers which in modern terms would give rise to a negative result were not considered possible.

Activity 10.3, page 126

1 You can set out the problem as
$$-5x + 6y + 8z = 290$$
$$3x - 9y + 3z = 0$$
$$-13x + 2y + 5z = -580$$
where x, y and z are the prices of cows, sheep and pigs respectively. First divide the second equation through by 3, and then re-order the equations to give
$$x - 3y + z = 0$$
$$-5x + 6y + 8z = 290$$
$$-13x + 2y + 5z = -580$$

You can now 'eliminate' x from the second and third equations using the first one. This gives
$$-9y + 13z = 290$$
$$-37y + 18z = -580$$

You can solve these for y and z, and then substitute these values into the first of the re-ordered equations to obtain a value for x.

The prices of the various livestock are as follows: cows 70 pieces, sheep 40 pieces and pigs 50 pieces.

Activity 10.5, page 127

1

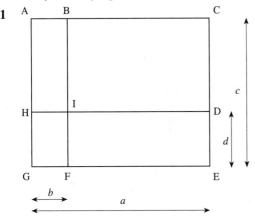

The area of the rectangle BCDI with sides of lengths $(a-b)$ and $(c-d)$ is $(a-b)(c-d)$.

From the diagram, you can see that this area is the same as the area of the large rectangle ACEG with sides of lengths a and c provided that we subtract off the area of the L-shaped border, which consists of the rectangles ABIH, IDEF, HIFG.

The L-shaped border is made up of two rectangles ABFG and HDEG of areas bc and ad respectively. However, if you subtract both these from the area ac then the area of the small rectangle with area bd in the bottom left hand corner (that is, the area of HIFG) has been subtracted twice when really it should only be subtracted once. Therefore the area of HIFG has to be added in again once to compensate. If you write this out in the form of an equation, you get $(a-b)(c-d) = ac - bc - ad + bd$.

Activity 10.6, page 128

1 a By the construction, the lengths of the lines EF, EH, FG, HG, AK, CL, AC, KL are all the same; call these lengths b. Also the lengths ED, DB, AB, AE are the same, call these lengths a. Then the lengths GI, BI, CB, CG are all $a-b$. This is a subtracted number since $a > b$.

The proposition explains that:
area ABDE + area ACLK =
        area HGLK + area EDIH + area GIBC

Using modern notation this is
$a^2 + b^2 = ab + ab + (a-b)^2$. By re-arranging this you get $(a-b)^2 = a^2 - 2ab + b^2$.

b This is a special case of al-Khwarizmi's equation in which c is taken to be the same as a and d the same as b.

Activity 10.7, page 129

1 In al-Khwarizmi's equation you can put a and c equal to zero giving $(-b)(-d) = 0 - 0 - 0 + bd = bd$ which provides a rule for multiplying negative numbers together.

2 You can obtain Brahmagupta's rules from al-Khwarizmi's equation by making the following substitutions in the equation.

Positive divided by positive is affirmative:

put b and d equal to zero and replace c by $\dfrac{1}{e}$ where e is a positive number.

Similarly, $(-b)\left(-\dfrac{1}{e}\right) = \dfrac{b}{e}$, $(a)\left(-\dfrac{1}{e}\right) = \dfrac{-a}{e}$ and $(-b)\left(\dfrac{1}{e}\right) = \dfrac{-b}{e}$.

Activity 10.8, page 129

1 One possibility is that during the intervening period the development of calculus occurred and that all the associated disputes occupied European mathematicians.

Activity 10.9, page 130

1 This is an interesting passage and it is worth noting various points about it.

● Although written in 1673 Wallis's text is very easy to read today and the interpretation seems up to date.

● Wallis provides a physical interpretation of negative numbers, and thus a reason to consider their existence and their validity as solutions of equations.

● Wallis proposes that the positive and negative numbers provide a useful way of modelling directions along a straight line with the magnitude of the number providing the distance.

● Wallis's way of visualising operations on numbers by using movements on the number line has similarities with Activity 10.1.

Activity 10.10, page 131

1 a You can find the translation of Arnauld's extract in the hints to this chapter.

b Arnauld is using the ordering of numbers along the number line and assumes that if a number is on the left of another on the number line then their order is preserved when each of these numbers is multiplied by a third. So that if a is on the left of b then he assumes that $a \times c$ is on the left of $b \times c$; that is, if $a < b$, then $ac < bc$.

c The above property is true if you multiply by a positive quantity. However you know that this is not true of multiplication by negative numbers when the inequality must be reversed. So if $a < b$ and we multiply both sides of the inequality by $c < 0$, then $ac > bc$. You know that you have to reverse the inequality when you multiply by a negative number. This means that there is no contradiction in the rules of multiplication when they are applied to negative numbers as Arnauld suggests.

Activity 10.11, page 131

1 a Lists of assumptions

Euler

● assumes that the outcome of multiplying together two numbers is not affected by the order in which we take the two numbers
● models a negative number as a debt and uses obvious properties of debts to infer properties of negative numbers
● assumes that the product of two negative numbers must be either a positive or a negative number, that is, it cannot be something new
● assumes that if you take a number and in turn multiply it by different numbers then the outcome cannot be the same in both cases
● assumes that when multiplying two numbers together it is possible to calculate the size (absolute value) of the product by multiplying together the sizes (absolute values) of the individual numbers.

Saunderson

● assumes that you can establish that numbers are in arithmetic progression if there are three or more of them to consider
● assumes that if you know the first two terms of an arithmetic progression, then the third can easily be found
● assumes that if all the terms in an arithmetic progression are multiplied by the same factor then the resultant terms (taken in the same order) will also form an arithmetic progression
● assumes that, even though you might start with an arithmetic progression with decreasing terms, having multiplied each by the same factor then the outcome may be a progression with increasing terms, or vice versa.

b Arnauld's objections to negative numbers are summarised below to inform the debate. Arnauld's objection relates fundamentally to ratios of numbers. He believes that 1 is to a what b is to ab. In other words he believes that $1 : a$ is equivalent to $b : ab$. Examples are:
$1 : 3$ is equivalent to $4 : 12$;

$1 : \frac{1}{3}$ is equivalent to $\frac{1}{4} : \frac{1}{12}$.

But he does not believe that $1 : -4$ is equivalent to $-5 : 20$.

The reason he gives is that 1 and -4 cannot be in the same ratio as -5 and 20 because $1 > -4$ whereas $-5 < 20$. On that basis alone he disbelieves the rule for multiplying negative numbers together.

How might a debate between Euler and Arnauld have been played out on the basis of the extracts? It all depends on the different assumptions that each makes about operations of numbers.

Presumably Arnauld would be prepared to accept that $1 : -3$ is equivalent to $4 : -12$ since his ordering principle applies. In a debate Arnauld could be persuaded to accept that $-b$ multiplied by a (where a and b are both positive) produces a negative result. However he would not accept that $1 : 4$ is equivalent to $-3 : -12$ since this violates the ordering principle, and thus presumably would not accept that a multiplied by $-b$ is valid.

This logically means that Arnauld cannot accept that a times $-b$ is the same as $-b$ times a.

On the other hand Euler assumes that the order in which you multiply a positive by a negative number does not affect the outcome. A debate between Euler and Arnauld could centre on this difference of opinion with one trying to convince the other of his view. Euler could be made to justify his assumption. This might be the chink in Euler's or Arnauld's armour.

Saunderson and Arnauld have more obvious differences of opinion. Saunderson considers numbers in arithmetic progression; for example, the decreasing sequence 3, 0, -3. By multiplying each of the numbers in the sequence by -4 say, he assumes we get another sequence which starts with -12, 0. The third member of the sequence must be 12 (using the arithmetic sequence properties) and this led him to deduce that -3 multiplied by -4 is 12. Arnauld would clearly object to this on the grounds that Saunderson had started with a decreasing sequence and yet ended up after multiplication with an increasing sequence. The ordering principle had been violated.

At the end of the day, the three mathematicians make different assumptions; perhaps Euler and Saunderson are prepared to go further than Arnauld and that is where the fundamental difference lies.

Activity 11.1, page 138

1 You may have established the result for the product of two negative numbers by putting equal to zero the first number in the pair forming each of the subtracted numbers. Yet you have no idea whether this rule is valid or not since you have not proved it as such. Equally you have no way of knowing whether your equation is valid with a and c both put as zero; the original argument was a geometric one with a and c representing lengths of the sides of a rectangle. It makes no sense geometrically to put these lengths equal to zero and to retain the existence of a rectangle.

Activity 11.2, page 138

1 a If you take $+2$ as representing two steps to the right, then $-(+2)$ represents two steps to the left since the minus sign indicates a change of direction.

You can consider $-(+2)$ to be the same as $(-1) \times (+2)$. Hence $(-1) \times (+2)$ represents two steps to the left, that is, the number -2.

If the order in which the multiplication is carried out is reversed, then the result is the same although the analogy is slightly different. You start with -1, a single step to the left, and then you multiply it by 2. This gives two steps to the left, that is the number -2.

This extends to any positive number a, so you have shown that $(-1) \times (a) = (a) \times (-1)$. Now multiply the result of the first case by 3; this produces six steps to the left. Hence $(+3) \times (-1) \times (+2)$ is -6. But you have already decided that $(+3) \times (-1)$ should be the same as (-3). So we have shown that $(-3) \times (+2) = -6$. A similar argument can be applied using the step model to any negative number $-b$ and positive number a.

By analogy with the step model the product of a negative number with a positive number (with the negative one as the first number in the product) is negative.

b A similar reasoning applies to the product of two negative numbers. In the first instance you start with -2 which represents two steps to the left. So $-(-2)$ represents a change of direction applied to this step and gives two steps to the right. Thus the operation $(-1) \times (-2)$ can be considered to be the same as the number $+2$. The rest of the explanation follows in an analogous way to part **a**.

Activity 11.3, page 139

1 a If i is a line of unit length, inclined at 90° to the positive number $+1$, then you can consider multiplying by i as applying a rotation of 90° anti-clockwise. Since $i^2 = i \times i$, multiplying by i^2 is the same as multiplying by i twice, or rotating by 90° twice. This is the same as rotating by 180° or multiplying by -1.

b

×	+1	−1	+i	−i
+1	+1	−1	+i	−i
−1	−1	+1	−i	+i
+i	+i	−i	−1	+1
−i	−i	+i	+1	−1

2 a $(2+x)(3-x) = 6 + 3x - 2x - x^2$
$= 6 + x - x^2$. By analogy
$(2+i)(3-i) = 6 + 3i - 2i - i^2$
$$= 6 + 3i - 2i + 1$$
$$= (6+1) + (3-2)i$$
$$= 7 + i$$

b $(2+i)(3.5 - 27i) = 34 - 50.5i$,
$(2+i)(0 + 94i) = 188i - 94$
$(3.5 - 27i)(0 + 94i) = 2538 + 329i$

c $(2+i) + (3.5 - 27i) = 5.5 - 26i$,
$(2+i) + (0 + 94i) = 2 + 95i$
$(3.5 - 27i) + (0 + 94i) = 3.5 + 67i$

Activity 11.4, page 141

1 One counter-example is given in the answer to question **1c** in Activity 10.10, that is, if you have two positive numbers a and b which satisfy $a < b$ and multiply both sides of the inequality by $c > 0$, then $ac < bc$. However if you multiply by $c < 0$ then you have to reverse the inequality to give $ac > bc$.

A second counter-example is: if a and b are positive numbers then their sum is greater than each of the individual numbers; this is not true if a and b are negative numbers.

Activity 11.5, page 141

1 a The required number under addition is 0 since $a + 0 = 0 + a = a$ that is, adding zero to any number in either order leaves that number unchanged. Zero is called the identity under addition. The number which plays a similar role in multiplication, that is, the identity under multiplication, is $+1$. You can justify this because $a \times 1 = 1 \times a = a$.

14 Answers

b The required number is -2 as $2+(-2)=0$. The same is true if the addition were carried out in the reverse order. For this reason -2 is called the inverse of 2 under addition since under addition 2 and -2 combine to give the identity under addition. All numbers have an inverse under addition; given any number, it is always possible to find another which will combine with it under addition to give 0.

c The number which is the inverse of 2 under multiplication is $\frac{1}{2}$ since $2 \times \frac{1}{2} = \frac{1}{2} \times 2 = 1$. It is not always possible to find an inverse under multiplication for every number. Take 0 for example: there is no number which when multiplied with 0 gives the identity under multiplication.

d The statement is not true because 0 does not have an inverse under multiplication. It would be true if it were modified to the following.

If a **non-zero** number is chosen at random, then the axioms require that it is always possible to find another **non-zero** number which, when combined with the original number using the operation $\times$, gives 1.

Activity 11.6, page 142

1 a The number 0 is an identity under f since it satisfies $f(a,0) = f(0,a) = a$.

The number 1 is an identity under g since for any a it satisfies $g(a,1) = g(1,a) = a$.

For every number a there is another number, $-a$, which combines with it under f so that $f(a,-a) = f(-a,a) = 0$.

For every non-zero number a there is another number, $\dfrac{1}{a}$, which combines with it under g so that

$$g\left(a,\frac{1}{a}\right) = g\left(\frac{1}{a},a\right) = 1.$$

b It is true for f since the order in which we add numbers together does not affect the outcome. For example, $2+3$ is the same as $3+2$. This is true in general and as a result addition of numbers is said to be a commutative operation.

The statement is true for g too since multiplication of numbers is a commutative operation.

Activity 11.7, page 143

1 Complex numbers $a+ib$ consist of a real part a and an imaginary part b. The number b can be zero so that the imaginary part can be zero. So all real numbers are at the same time complex numbers, they are special complex numbers in the sense that the imaginary part is zero. Zero, the identity under addition, is thus a perfectly valid complex number which performs the same function in complex numbers as it does to real; that is, when zero is added to any complex number it leaves that complex number unchanged. For example:

$(2+3i)+(0+0i) = 2+3i$. $(0+0i)$, or just plain 0, is the identity under addition.

In a similar way $(1+0i)$, or 1, is the identity under multiplication.

2 The inverse under addition of $2+i$ is $(-2-i)$ since $(-2-i)+(2+i) = 0+0i = 0$.

The inverse under multiplication is a bit trickier to find. You are trying to find a complex number which when multiplied by $2+i$ gives $(1+0i)$. Suppose that the inverse is $a+ib$ then

$(2+i)(a+ib) = 2a + ia + 2ib + i^2b$

$$= (2a-b)+(a+2b)i$$

This should be $(1+0i)$ and so you know that $2a-b=1$ and $a+2b=0$. You can solve these simultaneously to find that $a = \frac{2}{5}, b = -\frac{1}{5}$.

Activity 11.9, page 144

1 a To build up a step of length 12 to the left from the basic building blocks: first take a step of length 1 to the right, increase its length by multiplying by 12 and then reverse the direction by incorporating a minus sign in front which is effectively multiplying by -1. Thus $1 \to 12 \to -12$.

b When tripled, a step of 2 to the left becomes a step of 6 to the left; reversing produces a step of 6 to the right; then doubling produces a step of 12 to the right.

Using Clifford's notation, this is $k2(r3(-2)) = +12$, or in usual notation $2(-3)(-2) = +12$.

The same outcome can be achieved by for example taking the step of the 2 to the left, reversing it, then tripling and then doubling. The order of the tripling and doubling operations can be interchanged and they can take place before or after the reversal. The reversal operation commutes with the other operations and they commute with each other, that is, the order in which they are carried out does not affect the outcome.

2 a To do this you take the standard step and reverse it and then multiply it by one-half, or vice versa.

b Performing this step three successive times gives a step of one and a half times the standard length and with direction to the left. Reversing gives a step of one and a half times the standard length and with direction to the right. It can be expressed in symbols as

$$-\left(\left(-\tfrac{1}{2}\right)+\left(-\tfrac{1}{2}\right)+\left(-\tfrac{1}{2}\right)\right).$$

Activity 11.10, page 145

1 a You take a step of one unit to the right (that is labelled by **i**) and multiply it by the scalar 2. This gives you a step of length two units to the right. The incorporation of the minus sign (equivalent to multiplying by the scalar -1) reverses the direction. Thus $-2 \times \mathbf{i}$ represents a step of 2 units to the left.

b $\mathbf{i} + \mathbf{j}$ represents a step of length $\sqrt{2}$ inclined at an angle of $45°$ to each of the basic steps **i** and **j**. You will find the details in the hints to this activity.

c You should consider $-\left(\tfrac{1}{3} \times \mathbf{i}\right)$ and $-\left(\tfrac{3}{2} \times \mathbf{j}\right)$ separately first. $\left(\tfrac{1}{3} \times \mathbf{i}\right)$ represents a step to the right with size one third of the size of the basic step. Then incorporating the minus sign changes the direction so $-\left(\tfrac{1}{3} \times \mathbf{i}\right)$ represents a step of one-third to the left.

$\left(\tfrac{3}{2} \times \mathbf{j}\right)$ represents a step upwards with size one and a half times the size of the basic step. Then incorporating the minus sign changes the direction so $-\left(\tfrac{3}{2} \times \mathbf{j}\right)$ represents a step of one and a half downwards.

The combination of the two produces a step as shown in the figure below.

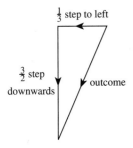

Activity 11.11, page 145

1 a $I \times I \times \mathbf{i}$ represents the application of the turning operation I, not once but twice, to the step **i**. Applying I once to **i** gives **j** as outcome. If you apply I again then you must apply it to the outcome **j** to obtain $-\mathbf{i}$. So $I \times I \times \mathbf{i} = -\mathbf{i}$.

b In a similar way, $I \times I \times \mathbf{j} = I \times -\mathbf{i} = -\mathbf{j}$.

c You can express the results of parts **a** and **b** as

$I^2 \times \mathbf{i} = -\mathbf{i}$ and $I^2 \times \mathbf{j} = -\mathbf{j}$. Therefore I^2 has the effect of incorporating a minus sign into the basic steps and so it behaves in the same way as the scalar -1.

2 In each case you multiply out the brackets, and then you treat each subsequent term separately.
$(1 + I) \times \mathbf{i} = 1 \times \mathbf{i} + (I \times \mathbf{i}) = \mathbf{i} + \mathbf{j}.$

In Activity 11.10 you showed that $\mathbf{i} + \mathbf{j}$ is represented by a step of length $\sqrt{2}$ inclined at an angle of $45°$ to **i**.

$$(1 + I) \times \mathbf{j} = 1 \times \mathbf{j} + (I \times \mathbf{j}) = \mathbf{j} + (-\mathbf{i}) = \mathbf{j} - \mathbf{i}$$

The outcome here is the composite step $\mathbf{j} - \mathbf{i}$. You can show that this is a composite step of length $\sqrt{2}$ inclined at an angle of $45°$ to each of $-\mathbf{i}$ and **j**. In each case you can see that the effect of applying the operator $(1 + I)$ to the basic steps is to turn the step through an angle of $45°$ in an anti-clockwise direction. $(1 + I)$ is a turning operator.

$$(2 - I) \times \mathbf{i} = 2 \times \mathbf{i} - (I \times \mathbf{i}) = 2 \times \mathbf{i} - \mathbf{j}$$

$$(2 - I) \times \mathbf{j} = 2 \times \mathbf{j} - (I \times \mathbf{j}) = 2 \times \mathbf{j} - (-\mathbf{i}) = 2 \times \mathbf{j} + \mathbf{i}$$

In these two examples the operator $(2 - I)$ is acting on the basic steps **i** and **j** in turn. You can see the effect of the operator $(2 - I)$ on each of the basic steps by drawing diagrams showing both the outcome and the step with which you started. These are shown in the figure below. From these you can deduce that $(2 - I)$ is a composite operator consisting of both a turning, or rotation, and an enlargement. In each case the step with which you started has its size enlarged by a multiple of $\sqrt{5}$ and at the same time it is turned through an angle θ in a clockwise direction where $\tan \theta = 0.5$. The scalar term in the composite operator represents the scaling factor (in this case an enlargement) and the scaling, either a stretch or a contraction, will take place parallel to the step to which the operator is applied.

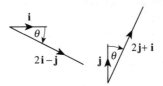

3 You can think of $2 \times I$ as being the same as $I + I$, so $(2 \times I) \times \mathbf{i} = (I + I) \times \mathbf{i} = \mathbf{j} + \mathbf{j} = 2 \times \mathbf{j}$. Thus the effect of multiplying the operator by a number is to multiply the original outcome by that same number. In this case since the number is greater than 1 a stretch as been applied upwards. So the scalar factor multiplying the turning operator has the effect of applying a stretch or a contraction at right angles to the step to which the operator was applied.

Taking the same approach with the other example, you obtain in a similar way $\left(1+\left(\frac{2}{3}\times I\right)\right)\times i = i + \frac{2}{3}j$. This is a composite turning and scaling operator similar to that in question **2.** However in question **2** the scaling factor produced a stretch or contraction parallel to the step to which it was applied. In this case the stretching or contraction occurs at right angles to the basic step to which the operator is being applied.

Summarising questions **2** and **3**, if the step itself acquires a scalar factor (as in question **2**) then the stretch/ contraction is parallel; if the turning operator has the scalar factor applied (as in question **3**) then the stretch/ contraction is at right angles.

Activity 11.12, page 146

1 a

$$a \times b = (2i - 3j) \times (-i + 5j)$$
$$= -2(i \times i) + 3(j \times i) + 10(i \times j) - 15(j \times j)$$
$$= -2 + 3(j \times i) + 10(i \times j) - 15$$
$$= -17 + 10(i \times j) + 3(j \times i)$$

2 a Taking the equation $(I \times i) = j$ and multiplying both sides on the right by **i** gives $I \times i \times i = j \times i$.

But $i \times i = 1$ so that $I = j \times i$.

Similarly by starting with the equation $(I \times j) = -i$ and multiplying both sides on the right by **j**, you obtain $I \times j \times j = -i \times j$.

Also $j \times j = 1$ so that $I = -i \times j$.

b In the first part of **a** you have taken $(I \times i)$ and multiplied it on the right by **i** to give $(I \times i) \times i$. However, subsequently you evaluate this as $I \times (i \times i)$ since you use $i \times i = 1$. The placing of the brackets in these expressions is important. You have assumed that the brackets can be taken away or re-positioned without having any affect on the result. In particular you have assumed that $(I \times i) \times i = I \times (i \times i)$. Such an expression would be true of real numbers, and you are assuming it is true in this context also.

c This follows very easily from part **a** since $I = j \times i$ and $I = -i \times j$.

3
$$a \times b = ((-6 \times i) - j) \times (-i - j)$$
$$= 6(i \times i) + (j \times i) + 6(i \times j) + (j \times j)$$
$$= 6 - (i \times j) + 6(i \times j) + 1$$
$$= 7 + (5 \times i \times j)$$
$$b \times a = (-i - j) \times ((-6 \times i) - j)$$
$$= 6(i \times i) + 6(j \times i) + (i \times j) + (j \times j)$$
$$= 6 - 6(i \times j) + (i \times j) + 1$$
$$= 7 - (5 \times i \times j)$$

Activity 11.13, page 147

1 These investigations are left to you.

Activity 11.14, page 147

1 Applying the turning operator J to **j** gives **k** and applying it to **k** gives $-j$. Thus $J \times j = k$ and $J \times k = -j$. From these we can deduce that $J \times j \times j = k \times j$ and $J \times k \times k = -j \times k$ so that $J = k \times j = -j \times k$.

The other similar equations are $K \times k = i$, $K \times i = -k$, $I \times i = j$, $I \times j = -i$, $K = i \times k = -k \times i$ and $I = j \times i = -i \times j$.

2 a If you apply J again to $J \times j = k$ then you obtain $J \times J \times j = J \times k = -j$. Hence $J \times J = -1$ and similarly for the other turning operators.

b
$$I \times J \times K = (j \times i) \times (k \times j) \times (i \times k)$$
$$= (j \times i) \times (-(j \times k)) \times (-(k \times i))$$
$$= (j \times i) \times j \times (k \times k) \times i$$
$$= (j \times i) \times j \times i$$
$$= -(i \times j) \times j \times i$$
$$= -i \times (j \times j) \times i$$
$$= -i \times i$$
$$= -1$$

c
$$I \times J = (j \times i) \times (k \times j)$$
$$= -(i \times j) \times (-(j \times k))$$
$$= i \times (j \times j) \times k$$
$$= i \times k$$
$$= K$$

and similarly for the other combinations.

Activity 11.15, page 148

1 No answers are given to this activity.

Practice exercises

Chapter 2, page 151

1 **a** 42.42638889, 1.414212963

b $\sqrt{2}$

c The answer needs to be a diagram like Figure 2.11.

d **Input** N, A

    repeat

$$\tfrac{1}{2}\left(A+\frac{N}{A}\right)\to A$$

    display A

    until satisfied

    Output The square root of N.

e 2, 2.75, 2.647727273, 2.645752048, 2.645751311, 2.645751311

2 Babylonians used the result in specific cases. Pythagoras's theorem involves proof, and there is no evidence that the Babylonians proved the result.

3 1; 0,45 and 0; 59,15,33,20

4 **a** How much grain is left over if as many men as possible are given 7 sila from a silo of grain.

b 40,0 times 8,0 to give the number of sila in a silo.

c To divide by 7, multiply by the reciprocal of 7.

d By multiplying

e The reciprocal of 7 has been taken as 0; 8,33 instead of 0; 8,34,17,8

5 $x=\sqrt{\left(\dfrac{a}{2}\right)^{2}+b}-\dfrac{a}{2}$

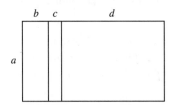

Chapter 3, page 152

1 **a** Let AB intersect l at the point K. Draw a circle centre K, radius KA, to cut l at C, and a circle centre K, radius KB, to cut l at D. Then $KA+KB=KC+KD$ so $AB=CD$.

b Let AB produced intersect l at the point K. Draw circles as for part **a**. Then $KA-KB=KC-KD$ so $AB=CD$.

c Take any point C on l and draw CA. Construct an equilateral triangle on CA, with the third point called E. Draw a circle centre A, radius AB, to cut EA at the point F. Draw a circle centre E, radius EF, to cut CE at the point G. Draw a circle centre C, radius CG, to cut l at the point D. Then $CD=AB$.

2 Construct an equilateral triangle ABC on AB. Draw a circle, centre B, with radius BP, to cut BC at D. Draw a circle, centre C, with radius CD to cut AC at E. Draw a circle, centre A, with radius AE, to cut AB at Q. Then $AQ=BP$.

3 First construct a rectangle twice the area by constructing a rectangle with a common base and the same height as the triangle. Then bisect the base of the rectangle to give a new rectangle with the correct area. Call this rectangle GHIJ, with HI as one of the shorter sides. Draw an arc of a circle, centre H, with radius HI. Extend GH to intersect this arc outside GH at N. Find the mid-point of GN and construct a semi-circle with this point as centre. Construct a perpendicular to GN at H. Call the point of intersection of the perpendicular with the semi-circle K and construct the required square on HK.

Chapter 4, page 153

1 **a** Diagram of the following form.

b $a(b+c+d)=ab+ac+ad$
area of large rectangle = sum of areas of small rectangles.

2 Using the algorithm on a graphics calculator, the greatest common divisor of 2689 and 4001 is 1.

3 If a measures b then $ka=b$ for some integer k. Hence $k^{2}a^{2}=b^{2}$, so a^{2} measures b^{2}.

4 Archimedes: spirals, measurement of a circle, or approximating pi, quadrature of a parabola, trisecting angles, measurement of area and volume

Chapter 5, page 154

1 **a** $x^{2}+10x=39$

b

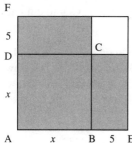

c $(x+5)^2 - 25 = 39$

$$(x+5)^2 = 64$$
$$x+5 = 8$$
$$x = 3$$

2 Al-Khwarizmi was only interested in quadratic equations with positive coefficients; the case in which 'squares and roots and numbers equal zero' has no real solutions.

3 Western Arabic numerals. The symbols for one, two, three and nine are very alike in the east and west. In the western Arabic versions, the four and five are too similar to each other for clarity. In the eastern version, it is the symbols for four and six, and for seven and eight which are too similar.

4 Halving the number of coefficients makes 6; 6 multiplied by itself makes 36. Adding 64 to 36 makes 100; the root of 100 is 10. Subtracting 6, the answer is 4.

5 $y = x - 200$

$$y^2 + 407y = 19350$$
$$z = y - 40$$
$$z^2 + 487z = 1470$$
$$z = 3$$

The solution is thus $x = 243$.

6 Omar Khayyam: classification of equations, methods for solving quadratic equations, geometric method for solving cubic equations.

Chapter 8, page 155

2 $x = 2$. If you substitute x^2 for y in the equation for the circle, you discover that the points of intersection satisfy $x^4 + 4x^2 - 16x = 0$; and if $x \neq 0$, this is equivalent to the required cubic equation.

3

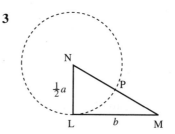

The required result is PM.

Chapter 9, page 156

1 Leibniz: development of calculus, recognising the relationship between integration and differentiation, development of a usable notation for calculus, development of a calculating machine.

2 Classes of problems: finding extreme values of curves (maxima and minima), finding tangents, and finding areas enclosed by curves, (quadrature). Calculus provides a universal method for the solution of all problems of this kind, rather than each example requiring its own particular or special approach.

Chapter 11, page 157

1 **a** Descending

b The value of $\dfrac{1}{n}$ becomes infinitely large.

c As n becomes close to zero and negative, $\dfrac{1}{n}$ becomes very large and negative. Wallis possibly thinks that the value of $\dfrac{1}{n}$ carries on increasing as n decreases.

d Wallis does not seem to understand that $\dfrac{1}{n}$ is not defined for $n = 0$, and that there is a discontinuity in the graph of $\dfrac{1}{n}$.

2 **a** $(56 + x) = 2(29 + x)$, so $x = -2$.
b It is negative.
c The father was twice as old as the son was 2 years ago.
d How old was the father when he was twice as old as the son?

Index

Text and illustration credits

We are grateful to the following for permission to reproduce copyright material:

Chelsea Publishing Company, Inc. for an extract from *The Mathematical Work of John Wallis* by J F Scott; Dover Publications, for extracts from *Elements, from Euclid, the Thirteen Books of the Elements*, 2nd edn. unabridged, by Sir Thomas Heath, *La Géometrie* by Descartes, 1954 and *The Geometry of René Descartes*, 1954; McGraw Hill Publishers, for an extract from *A Source Book in Mathematics* by D E Smith; Merlin Press Ltd for extracts from *The Life of Henry Brulard*, by Stendahl, translated by Jean Stewart and BCJG Knight; Open University Press, for extracts from *The History of Mathematics: A Reader* by J Fauvel and J Gray; Oxford University Press, for an extract from *Mathematical Thought from Ancient to Modern* by Kline; Routledge, for extracts from *Early Mesopotamia* by J N Postgate; J Wiley & Sons Inc. for an extract from *A History of Mathematics*, 2nd edn.; University of Wisconsin Press, Madison, Wisconsin for an extract from *Kushyar Ibn Labban, Principles of Hindu Reckoning*, edited and translated by Martin Levey and Marvin Petruck.

We are grateful to the following for permission to reproduce photographs and other copyright material:

Figure 1.2, Musée du Louvre; Figures 1.3a and 2.1, Karl Menninger, *Number Words and Number Symbols*, Massachusetts Institute of Technology 1969, p.163; Figures 1.3b and 1.8, O Neugebauer, *Vorleusugen Uber Geschichte det Anbiken Mathematischen Wissenschaften (Erster Band)* 1969, p.51, Springer-Verlag GmbH & Co. KG; Figure 1.4, Georges Ifrah, *De Wereld van Het Getal*, 1988, p.132, Servire. © 1985 Editions Robert Laffont S.A., Paris; Figure 1.5, Denise Schamandt-Besserat, *Archaeological Magazine*, Archaeological Institute of America; Figure 1.6, Lauros-Giraudon; Figure 1.7, The Trustees of the British Museum; Figure 2.3, Prof. A H Aaboe, *Episodes from the Early History of Mathematics*,

Random House, 1964; Figure 2.4a, E M Bruins & M Rutten, *Mémoires de la Mission Archéologique en Iran, Tome XXXIV, Textes Mathematiques de Suse*, 1961, Plate 6 Tablet K, Paris Librairie Orientaliste Paul Geuthner; Figures 2.4b and, 2.8b O Neugebauer, *The Exact Sciences in Antiquity*, 1969, Dover Publications/Constable & Co; Figure 2.6, we are unable to trace the copyright holder of this figure and would appreciate receiving any information that would enable us to do so; Figure 2.7, Lucas N H Bunt, Phillip S Jones and Jack D Bedient, *The Historical Roots of Elementary Mathematics*, Dover Publications/Constable & Co; Figure 2.8a, Yale Babylonian Collection; Figure 2.10, John Fauvel, *Mathematics through History*, QED Books, York; Figure 2.12, O Neugebauer and A Sachs, *Mathematical Cuneiform Texts*, 1945, (American Oriental Series vol.29). American Oriental Society, American School of Oriental Research, New Haven, Connecticut (we are unable to trace the copyright holder of this figure and would appreciate receiving any information that would enable us to do so); Figure 2.13, E M Bruins & M Rutten, *Mémoires de la Mission Archéologique en Iran, Tome XXXIV, Textes Mathematiques de Suse*, 1961, Plate 1 p. 23, Paris Librairie Orientaliste Paul Geuthner; Figure 2.14, O Neugebauer, *Mathematische Keilscriftexttexte 3. band, dritter teil, Verlag von Julius Springer*, 1935/1937. (Springer-Verlag GmbH & Co. KG); Figure 5.1, George Gheverghese Joseph, *The Crest of the Peacock: Non-European Roots of Mathematics*, 1992, Penguin Books; Page 73, *The Geometry of Rene Descartes*, title page, Dover Publications, inc; Figure 12.1, J Fauvel and J Gray, *History of Mathematics: A Reader*, The Macmillan Press Ltd. Adapted from O Neugebauer and A Sachs, *Mathematical Cuneiform Texts*, 1945, American Oriental Society, American School of Oriental Research, New Haven, Connecticut (we are unable to trace the original copyright holder of this figure and would appreciate receiving any information that would unable us to do so).